THE
BIOLOGICAL FARMER

A Complete Guide to the Sustainable & Profitable
Biological System of Farming

THE BIOLOGICAL FARMER

A Complete Guide to the Sustainable & Profitable
Biological System of Farming

Gary F. Zimmer

Acres U.S.A.
Austin, Texas

THE BIOLOGICAL FARMER

A Complete Guide to the Sustainable
& Profitable Biological System of Farming

Acres U.S.A., Publishers
P.O. Box 91299, Austin, Texas 78709 U.S.A.
phone (512) 892-4400 • fax (512) 892-4448
info@acresusa.com • www.acresusa.com

Printed in the United States.
Fourth Printing.

Publisher's Cataloging-in-Publication

Zimmer, Gary F.
        The biological farmer : a complete guide to the sustainable
& profitable biological system of farming / Gary F. Zimmer. — 1st ed.
        xiv, 354 p., 23 cm.
        Includes index.
        Library of Congress Catalog Card Number: 99-76577
        ISBN: 978-0-911311-62-4

        1. Organic farming. 2. Alternative agriculture. 3. Sustainable
agriculture. 4. Agricultural ecology. I. Title.

S605.5.Z56 2000   631.584
                                        QBI99-1664

CONTENTS

ACKNOWLEDGEMENTS

Any book requires a lot of work and in this case I received a lot of support and input from Bio-Ag staff workers, consultants, farmers and my family. Everyone who has touched my life in one way or another has contributed to this work. In particular, I would like to thank my wife, Rosemarie, for all the support, patience and editing she has provided to make this book possible. A big thank you, also, to my parents, Floyd and Anita Zimmer, for raising me on a dairy farm and teaching me work ethics and farm life values.

As a college student in the 1960s I learned to observe and question — to observe what was happening down on the farm and question agricultural practices. This biological method of farming could not have been put into practical and useable form without the testing and "thinking outside the box" of many farmers and researchers.

Much has been studied and written on the subject of biological farming; putting it all together in a working model was the challenge. Dr. Harold Willis, author of several great books on alternative agriculture, provided major assistance. Dr. Willis gathered and organized a lot of technical support information found in this book and he provided editing assistance on the practical sections. Thank you Harold, for without your driving force and assistance this book would never have been written.

ABOUT THE AUTHOR

Gary Zimmer was raised on a dairy farm in northeastern Wisconsin. He received training in dairy nutrition, earning a bachelors degree from the University of Wisconsin and a master's degree from the University of Hawaii (1970). After achieving his educational goals, he spent the next five years teaching a farm operation and management course at Winona Area Technical Institute, Winona, Minnesota. In addition to nutrition, he was required to teach courses in crops, soils and farm finance. His approach to teaching included outside speakers, on-farm projects, field trips and much study. He has always asked questions and looked at many ways of farming with an open mind. This led him to discover *The Albrecht Papers*, Brookside consulting training, work as a private consultant in southeastern Minnesota, and ultimately to promote a better understanding and appreciation of biological farming.

For the last 25 years, Gary Zimmer has been evaluating farming practices which has presented him with many interesting observations and methods for successful biological farming. In the fall of 1979, Gary and his family moved to Spring Green, Wisconsin, where, in the early 1980s, he and three partners started Midwestern Bio-Ag. With their main emphasis on education, Midwestern Bio-Ag currently works with more than 3,500 farms and has 75 trained consultants. Besides being president of Midwestern Bio-Ag, for the last five years, Gary has also managed the Bio-Ag Learning Center, a livestock and crop demonstration farm where he tests products and methods for successful biological farming. Otter

Creek Organic Farm is also operated and co-owned with his son Nicholas. The 240-acre farm is certified organic and grows edible beans, corn and forages. On the 100 acres of creek bottom pasture, rotational grazing is done either with dairy heifers or beef cattle. The first major test project on this farm was to see how rapidly low-fertility, run-down soils could be fixed with organic methods. Soil tests and a follow-up story are included in this manual.

Gary continues to be an educator at heart and is a much sought-after speaker. He conducts presentations from coast to coast for diverse groups of people, including farmers, farm financial groups, private farm consultants, environmental groups and other agribusinesses.

ABOUT THIS BOOK

This is a training manual for farmers and Bio-Ag consultants. If we are to change agriculture to a more sustainable biological system, we need to work together. If this book helps you become a better farmer or consultant, it has served its purpose.

Midwestern Bio-Ag's emphasis is on information, education and ideas for crop and livestock farmers. We make our living by providing high quality programs backed up by tested and proven products. These programs have their foundation in the "rules of biological farming."

First published in 1983, this book began as a training guide for Bio-Ag consultants. It has been added to and revised many times. In 1993 it was put into a ring binder and made available to our farmers with the belief that if farmers know and understand the *whys* and *hows* of biological farming, they can be more successful.

We have no secrets. We work with nature, feeding soil life, balancing soil minerals and tilling soils with a purpose. This is a practical how-to guide to give you ideas and suggestions for making farming more fun and profitable. Biological farming is a type of farming — another way — with beliefs, observations and rules to be followed. I will not try to convince you of the benefits of biological farming, but rather give reasons and explain how biological farming works. I'm sure there are other ways to farm biologically using other materials and methods, but the concept is the same.

No question about it, you can get high yields with lots of commercial fertilizers and synthetic chemicals for weed, insect and disease control. There are, however, some concerns and problems. Is this food production system safe and sustainable? At best we hope we are not creating more problems in the future. But we do have a lot of people to feed, and we do need to keep production up. Biological farming does not mean less production. It does mean working toward elimination of limiting factors to production. Once the soil's chemical (nutrients), physical and biological properties are in balance, you can expect optimum production, even in bad years.

We don't credit any one thing for making biological farming successful. Maybe it's the calcium source, maybe it's the change in how we till the soil, or the green manure crops we grow. Maybe it's the balanced, higher quality fertilizer. Maybe it's all of the above. Biological farmers manage the whole system and are not looking for miracles. They evaluate soils, crops and farming practices, and put together a whole-farm management system. That's what this *training manual* is designed to do — to give ideas and *guidelines* to follow for successful biological farming.

In 1995, Midwestern Bio-Ag consulted with Dr. Harold Willis to assist with gathering, organizing and editing technical and background material for the much-expanded book you now hold. Without him, this book would never have come to fruition. The author — and all the farmers and consultants who will benefit from this book — owe Dr. Willis a tremendous debt of gratitude.

This book has two objectives: one is to give the technical information — the *whys* — and the other is to present the practical *hows*. There is no one right way; this is not an exact science. But there are ways and practices that make sense. The principles of nature are the same for all: provide a soil which is well drained, loose and crumbly, with lots of soil life, and mineralized with all the essential elements — that's the objective. Select proper sources of materials and manage the excesses as well as the deficiencies. That is what it takes to grow high-yielding, healthy crops of any kind on all soils. Certain crops have different requirements which can be addressed with specific crop fertilizers added to a balanced soil. In other words, balance the soil minerals and use crop fertilizers for specific crops and conditions.

The "how-to," practical parts of the book are mixed in throughout. Many examples are given — not to be copied exactly, but to give ideas on how to make *any* farm work more efficiently and profitably.

Evaluate where the soils are as a starting point. Building healthy, mineralized soils takes time. The speed of correction has to do with soil types and present soil conditions. Hard, dead, heavy soils seem to take the longest to improve. The speed the farmer uses to correct mineral balance and follow healthy practices also affects the time it takes to improve the farm. Even with the implementation of new practices and major mineral corrections, it still takes time for the soil to change. For most farms, it takes three to five years, which is really a short time considering how long it has taken to get the land in the shape it is now. Certain observations and clues can be seen along the way toward correction. Feed quality and insect pressure usually change first.

After reading and studying the book, if you still have questions or want more information, please do contact our office. To change agriculture it will take a lot of model successful biological farms. There will also be a need for good consultants to put practical programs together and guide the farms toward success. If you have an interest in becoming either one of those, let us know. Midwestern Bio-Ag has available additional educational booklets and videos, and a complete consultant training program.

AN INTRODUCTION TO BIOLOGICAL FARMING

How can farming be profitable? With input costs soaring and returns declining, problems on the farm seem to be getting out of hand. The business of farming is very confusing, with such complex questions as: In whom do I trust for information and advice? What methods will work on my farm? What products should I buy? What level of spending is profitable? Can I make improvements? Do I have limiting factors? Is my method of farming harming the environment? Are my soils eroding? Is my method of farming sustainable for future generations? How much time will it take to get my soils to maximum health and production?

I recommend that you take a common-sense, basic approach to farming. There are ways to reduce your input costs and increase your profits, while at the same time improving your soil conditions and livestock health. I call it *biological farming*. It utilizes resources of both science and nature in a superior farming system. It works *with* natural laws, not against them, which is how things were designed to work. Widespread and serious man-made ecological destruction — poisoned air, water and food — tells us that methods which try to overpower nature will fail in the long run. Biological farming improves the environment, reduces erosion, reduces disease and insect pressure, and alters weed pressure, and it accomplishes this while working in harmony with nature.

What Makes a
Successful Biological Farmer?

Skilled biological farmers learn how to take care of soil life — they nurture it, feed it a balanced diet, and use tillage tools and methods to enhance soil life. Farmers must understand proper use of livestock manure, compost and green manure crops. They learn how to evaluate soil for its health, tilth and soil life. Biological farmers learn how to evaluate crops, roots and plant deficiencies as well as the *whys* and *hows* of insect, disease and weed indications. Finally, biological farmers develop an understanding of fertilizers and soil fertility, the steps and methods to get soils in balance, and the proper use of fertilizers. Biological farmers learn the "when" and "how" of fertilizer use for soil correction, feeding soil life, balancing nutrients and feeding the crop with the proper balance between soluble and slow-release materials.

Three important parts of your soil are (1) the organic particles that serve as a reservoir of plant foods, (2) the soil minerals, and (3) the living portion, consisting of bacteria, fungi, algae and larger organisms such as earthworms. These organisms are alive and need air, water, organic matter (food) and a safe place to live. Work with them, because the productivity of your farm is in direct proportion to the number, activity and balance of soil organisms.

Sustainability is a key factor in becoming a biological farmer. Important points in becoming a sustainable farmer are that soil erosion is stopped, soil tilth is ideal, soil nutrient balance is correct, and soil life is abundant (in many soils a good measure is a minimum of 25 earthworms per cubic foot of soil). The value in being sustainable is that it helps to maximize crop production with maintained soil fertility levels, while minimizing disease and insect problems within the limits of environmental conditions of proper moisture and degree growing days.

Balance is the key, not only for economic, but ecological reasons as well. It is essential to provide all elements to your crops and to soil organisms in the proper balance. An excess of some nutrients can be as limiting as deficiencies of others. Agronomists and soil scientists have written that at least 16 elements are needed to grow plants. *The productivity of a soil can never be greater than the plant food element in least supply.* Look at the table in Chapter 4 to see what's in the soil. You need to make these nutrients "exchangeable," or available to the roots of the plant. Because nutrients can interact, an excess of some elements can cause a shortage of others, even though it appears there is enough on a soil test. Managing your soil and crops to produce large root systems that will recover the nutrients, plus working with soil organisms so they make nutrients available and exchangeable, can make farming fun and profitable.

Questions often asked by farmers about biological concepts include:
- Do we have to add all of these 16 elements to the soil?
- Do we need to return the elements to the soil in the amount that we take off?
- What types of materials should we apply for greatest efficiency and return?
- How much should be applied for maximum return?
- Should we use all soluble materials (commercial fertilizers) or provide slow release with more natural, complete fertilizers?
- What are the benefits of soil nutrient balance?
- What about calcium, magnesium, sulfur and trace elements?
- Do I really need green manure crops?
- How do I best manage residues?
- What is the best way to handle livestock manure; do I have to compost it?
- Where do biologicals, growth stimulants and plant hormones fit in?
- Dare I use herbicides, insecticides and biotechnology on my farm?
- What about feed quality and livestock health? How do we measure them, and how do we get them?

In this manual, we will answer these and other questions about successful biological farming practices.

What's Wrong with the N-P-K Approach to Farming?

Does it make sense to use high levels of only highly concentrated water-soluble nutrients? The N-P-K-pH chemical approach to farming is both incomplete and wasteful.

Nitrogen — It is not a mathematical thing. Crop rotation, the nitrogen source used, and when and where we put the nitrogen, all have a bearing on how much nitrogen we need. Also soil air, soil life, organic matter, and the presence and balance of other elements (such as sulfur and calcium) have an effect on nitrogen use and efficiency. Biological farmers do not want to use any more nitrogen than absolutely necessary, not only because of cost and possible environmental pollution, but also because excess nitrogen suppresses long-term stable biological processes in the soil.

In a test at the University of Minnesota, when 200 pounds of nitrogen were added in one application, the yield was just four bushels of corn more than when 100 pounds were applied in split applications. What happens to the extra nitrogen you apply? Does it benefit the soil, the environment — or your water?

Phosphorus — With phosphorus, not only can highly soluble commercial fertilizers have a hardening effect on the soil, but only a small amount of the total phosphorus is available to the growing crop because most of it changes back into the insoluble rock phosphate form. Biological farmers believe higher levels of available soil phosphorus are beneficial for increasing plant health and feed quality. These levels are attained by using naturally mined rock phosphate, some high quality manufactured phosphorus, green manure crops, livestock manure, and biological activity.

Potassium — The most common commercial source of potassium used is muriate of potash, or potassium chloride, a strong salt containing 47 percent chloride. Research has found that chloride takes two years to leach below two feet in typical soil. While it is in the upper soil, high levels of chloride can kill beneficial soil life and injure roots. Far too high rates of potassium are often recommended, leading to soil imbalance and lower quality crops. Although high potassium levels often stimulate big piles of crops (yield), we can get high yields along with better quality (a better balance of nutrients in the food) by applying more calcium and less potassium to the soil.

Synthetic N-P-K fertilizers have few or no secondary and trace elements, yet all 16 elements are necessary for plant growth, not just the "big three" — N, P and K. Continued use of N-P-K alone can result in trace element deficiencies.

Does the N-P-K farming approach seem to be balanced with nature? Does it seem to be beneficial for soil life and crop health? How would your livestock perform if you treated them that way? It's the same thing. Look around: the soil gets harder and erodes, water quality deteriorates, animal and human health problems increase, pests and weeds proliferate, and more fertilizers and chemicals are continually needed. It's all related to how we treat "Mother Nature's" soil. Balance is the key, just as it is with livestock and human nutrition. We need a farming method that obtains "quality," and this means healthy, mineralized, balanced crops.

My introduction to biological farming came from teaching and working as a consultant. After growing up on a dairy farm, I spent eight years in college studying agriculture and obtained a master's degree in dairy nutrition. Then I taught agriculture for five years and finally returned to farming and working with farmers as a soils and nutrition consultant. That's when I came to the conclusion that *management, balance* and *efficiency* are what make farming profitable.

Biological farming will improve things. Don't just accept the farm situation the way it is. It can change and become profitable. Many farmers are doing it right now. There are many books and printed materials that provide useful information. Visit the farms, read the books, and check it out for yourself.

The purpose of this book is to give you an easy-to-follow guide so you can make intelligent decisions about soil fertility and soil management practices that are profitable and that make sense. It is designed to be a practical "how-to" guide.

Biological farming is a program, not just adding a single product and hoping for a miracle. It takes time to change things, and there is *no one single way to do things*. But there are ways that make sense. To be successful in the long run, you must use an approach where you are working *with* and properly utilizing nature's biological systems.

1

TWO MODELS: BIOLOGICAL & CONVENTIONAL

Some people seem to have a lot of trouble switching over to biological methods of agriculture. They can't seem to figure out how to go about things — what to do next. And their efforts often fail. But others take to biological farming easily. They can readily plan what to do and improvise new solutions to their own individual challenges. Why the difference?

How we know what we know. It really comes down to one's way of thinking. We all tend to operate on a base of knowledge and opinions that we have learned and absorbed from those around us throughout our lifetime. We often accept what others tell us simply because they have a couple of degrees behind their name. Or maybe we just do what everyone else we know is doing. We give in to peer pressure — we're afraid to be different.

We accept the "conventional wisdom" without ever questioning whether it is really right or whether there is a better way. But time and again, history has proved that the conventional wisdom is incomplete — or even totally wrong. Examples like the early belief that the world is flat or that sick people could be cured by bleeding them may seem ridiculous in the late 20th century, but some "facts" that we learned in school are already assigned to the dustbin of amusing misconceptions. "Knowledge" and "truth" are certainly imprecise and changeable things, as we limited humans try to discover them.

What is truth? There definitely are absolute truths, but how do we go about finding or recognizing them? The first step is to open your mind to the notion that you don't have it all figured out yet — and maybe the "experts" don't either. Second, you have to seriously consider with an open mind new ideas and possibilities — is there a better way? Third, you have to check it out for yourself — see what other people are doing, and eventually try it yourself.

What works? The real test of what is truth is what *works*. Does it produce right results — good results? Not only now, but next year — and next decade? Don't just consider what will benefit you or your bank account, but also what is good for humanity and our fragile environment. If you work closely with nature's systems, you can't fail.

Conventional agriculture. The system of agriculture that has become known as "conventional" (because almost everyone is doing it) has only been the conventional wisdom for about 50 years (since World War II). A few developments, like hybrid corn, have been around longer, but the heavy use of highly toxic chemicals and synthetic fertilizers is a post-war phenomenon. Your great grandfather didn't farm that way. But you may say, he didn't get the yields we get now. Today we have the tools and understanding that allow us to farm in an environmentally friendly way *and* get the yields *and* be the most profitable.

The initial results of what we call the conventional system of farming seemed good. Crop yields and animal production skyrocketed — we were awash in surpluses. But what about the long-term effects — the hard, dead soil, the poisoned groundwater, the increasing pest problems? And what about the "minor details"? — the fact that today's food is so lacking in vitamins and minerals that we have to give livestock costly ration supplements and take vitamin and mineral pills ourselves. Or the fact that most fruits and vegetables grown commercially have to be rushed to market before they spoil. What ever happened to *quality?*

Has conventional agriculture really fulfilled its promise? Does it really produce good results? Is there a better way?

Let's examine the basic philosophy that drives conventional agriculture. We could call the set of beliefs behind a certain practice a *model* or *paradigm*. The dictionary defines paradigm as a pattern, a model, or a principle. It seems clear to anyone who is willing to be objective that the basic philosophy behind conventional agricultural practices is that of *get*. In other words, man is superior to nature. He wants to *get* all the food and money he can. If weeds or bugs interfere, kill 'em! Our superior intellect has figured out that plants need only a handful of simple ions to grow. We can *get* more food by making our own fertilizers and giving plants an extra shot. And we can *get* even more by being more efficient, by farming bigger and faster.

Another system. But there is another paradigm, another viewpoint, another system of agriculture. It is more humble. It respects nature and realizes that man doesn't know it all. Plants and animals grow by natural laws, and they grow best when natural laws are followed, not overpowered. Life operates in natural cycles. One thing affects another. The way soil is fertilized controls the quality of crops. Crop quality affects animal and human health and productivity.

This system of agriculture relies on cooperation, working with nature, not against it. It could be called the way of *give*, not get. When we *give* the soil the right materials so an amazing array of soil organisms can do their jobs, we receive an abundance of nutritious, high quality food — without even thinking about getting. Certain natural organisms and mechanisms protect against diseases and pests — automatically. We only have to encourage them and get out of their way.

We call this paradigm *biological agriculture*, but others call it by different names. "Biological" emphasizes one of its pillars, life in the soil. Biological agriculture works with natural laws and systems and tries to help them operate more effectively. A healthy, balanced soil is the foundation, necessary for healthy plants and animals. Biological farming is not against using modern technology and new methods, but uses only those that do not interfere with natural systems and do not cause harm down the road.

The results of using biological methods are amazing. After a few years, soil structure improves, crop yield and quality are high, and animals are healthy and productive. Weed, disease and pest problems almost disappear. It really works. Farming is fun again.

Two systems of agriculture. We can summarize and compare conventional and biological systems of agriculture with the following table.

	Conventional	**Biological**
Basic outlook	Control nature. Maximize yield and profit. Short-term view.	Work with natural system. Increase health and quality. Long-term view.
Soil	Supports plant; supplies about a dozen elements.	A complex system: physical, chemical and biological factors.

	Conventional	Biological
Fertilizers	Synthetic, soluble salts emphasize N-P-K and pH. Replace what crop removes.	Natural or low-salt, some available and some slow-release. Balance all elements in soil. Also, feed the crop a balanced diet.
Crops	Often monoculture. Grow for market and yield.	Soil-building rotations. Grow for quality.
Weeds	Chemical control.	Manage soluble nutrients, mechanical control or spot herbicides; smother crops.
Pests/diseases	Chemical control, resistant varieties.	Natural control by good health, natural enemies and rotations.
Animals	Antibiotics. Push for production.	Probiotics. High production from quality feed, good health.
Economics	High inputs. Moderate profitability.	Low inputs. High profitability.
Environment	Chemical pollution, degraded soil with high erosion.	Little pollution. Good soil with low erosion.

Changing over. You can't simply switch from conventional methods to biological by just using some different fertilizers and mechanical weed control. You can't continue to use the same ways of thinking, because the two systems of agriculture have completely different philosophies and orientations. That's why some people have so much trouble changing over. They are still using conventional agriculture's goals and thought patterns — the old paradigm. The rules and goals can't be transferred.

You have to "reprogram" your mind to think in new ways, to seek new goals. It's not easy. You have to "unlearn" what you have done for decades. Question everything. Open your mind and eyes and look around. You have to imagine new possibilities, try new things, and be willing to be different from your neighbors.

But think about the results of the two systems. Are expensive chemicals, weeds and pests, sick animals and poisoned water really worth it? Wouldn't you rather have good soil, healthy crops and animals, and freedom from worry about poisons? Think about future generations. With conventional methods, can you give your children land fit to farm? Do you feel good about what you do? Are you proud of the way you farm? There is a better way — it has been proven on many farms around the country. The choice is yours.

A goal to work toward —
the model biological farm.
This is a system of farming
where the farmer thinks differently.

Objectives

- Soils: Get the soil alive with earthworms and other soil life, and balanced with minerals.
- Crops: Produce a quality crop which is a complete livestock nutrient source and which results in high yields.
- Livestock: Keep livestock comfortable, healthy and fed a balanced diet using high quality feeds with health-promoting "extras."
- Biological farming: Work with the systems of nature to develop a farm which is environmentally sound and which leaves the land, water, plants and animals in a healthy, productive state for all future generations.

Soils

The biological farmer . . .

- Understands that the soil is living — alive with many organisms and balanced minerals. Wants the soil in the best possible health. Knows that managing decay of organic matter, crop residues, and livestock and green manures are essential to his success; these are the feeds for soil life.
- Considers the effects on soil life, positive and negative, of all inputs and practices.
- Studies and understands nutrients. Knows when plants need them and that soil life makes some of them available to the roots. Would not add more fertilizer than is needed, but certainly wouldn't starve the plant.
- Uses sources of nutrients that are non-harmful to plant roots and soil life.
- Applies suitable products to correct soil nutrient imbalances.
- Knows that money invested to get and keep his soils in a healthy state is money well spent.

- Questions use of herbicides and really questions any use of insecticides. Knows he can farm using less or none with proper crop rotations and soil health.
- Finds cultural practices that fit his land and uses tillage to manage decay and control soil air and water.
- Knows less tillage is best, but some may be necessary.
- Knows shallow incorporation of residues is good for the soil life and speeds up decay.

Crops

The biological farmer . . .

- Knows that the quality of crops as complete nutrient sources is as important as yield. Does not sacrifice quality for a big harvest.
- Doesn't plant on dates; plants on conditions. Knows that early planting is not always best. Any condition less than ideal at planting is unacceptable.
- Evaluates crops and equipment, making sure plant stands are there and uniform.
- Fertilizes the crop for maximum health with a balance of nutrients in a form the plant can use on a continual basis.
- Knows that weeds, insects, soil conditions and plant growth (roots, tops, colors) are all tools and clues to evaluate his program.
- Knows the impact of crop rotation on reducing weed and insect pressure and improving soil, soil life and future crops.
- Does everything in his power to harvest at the correct time to maximize quality and yield.
- Provides proper storage to maintain crops as quality feeds. Includes inoculants on all fermented feeds to reduce spoilage.

Livestock

The biological farmer . . .

- Provides a clean, comfortable environment to keep livestock content and to promote maximum health and production.
- Feeds livestock a palatable, balanced, steady diet featuring quality home-grown feeds, fresh, clean water and access to free-choice minerals.
- Knows that feeds from a biological farming system are the best you can get. They are grown on healthy, living soils with balanced, adequate levels of minerals.
- Knows that these feeds do not always follow the "traditional" balance numbers. The rules may change.
- Frequently checks to be sure the feeds are in balance.

- Uses the best quality additions to balance nutrient deficiencies, including protein, energy and mineral supplements when necessary.
- Does not continuously feed animals antibiotics. Antibiotics are used as a treatment only when absolutely necessary to save the health/life of the animal.
- Does not inject animals with synthetic hormones in order to promote enhanced growth/production beyond the natural genetic ability.
- Understands and works with natural beneficial organisms within every animal.
- Feeds the extras — kelp, yeast, probiotics, digestive aids and extra vitamins – from day one.
- For cattle, feeds a ration of high quality forages (mineral-balanced from healthy soils), fed at high rates with lower levels of grain.
- For hogs and poultry, uses a ration of high quality grains (mineral-balanced from healthy soils) with low levels of forages.
- For breeding animals, knows that longevity is important, evaluates his livestock program based on animal appearance, health, comfort, manure, breeding efficiency and production.
- Knows that livestock must meet *all* of the above characteristics to be profitable.
- Treats the livestock manure to maintain nutrients and control odors (with natural phosphates, gypsum and beneficial organisms).
- Uses livestock manure as an important source of soil nutrients. Spreads manure in thin, even and timely applications.

Results
- The farming system is driven by profit, not just production.
- The farmer knows that to get all these factors working in tandem takes time. Conditions differ for every soil, farm and farmer.
- Profit in farming is generated more from the farmer's knowledge and management than from monetary inputs to the system.
- The farmer works on the cause of the problem. He does not want to always be putting out brush fires (dealing with symptoms). He knows that with biological farming, the problems will continue to decrease over time.
- The farmer understands nature and works with it rather than against it. He maximizes natural interactions of pests, predators and environmental conditions to his advantage on soils, crops and livestock.
- The farmer knows that healthy crops are not plagued by insects or diseases and can compete with weeds under proper management. He can spend less and worry less.

- The farmer knows that healthy livestock don't always get sick. They breed and produce when fed balanced diets from healthy, balanced soils. He can spend less and worry less.
- Biological farmers know that this is their best shot at being profitable, sustainable farmers. They can be proud of how they take care of their land, livestock and environment, and of the food they produce.
- Biological farmers know that farming can be profitable and fun.

2

HOW TO GET STARTED IN BIOLOGICAL FARMING

You may have heard that natural farming methods can give good results. You may be concerned about the hazards and pollution caused by toxic agricultural chemicals. You may feel uncomfortable using genetically modified plants. You may want to try out or change to biological farming methods, but you do not know how to go about it. With logic, guidelines and examples, I hope to provide you with answers and a program to follow.

What is Biological Farming?

You hear a lot of different terms today. *Sustainable, low-input, organic, alternative, biological, ecological, renewable, regenerative* and *natural* are some of them. Many of these terms are poorly defined, but they do have many things in common. What I call *biological agriculture* simply means farming with fewer chemicals, better soil stewardship, and a cleaner environment. It is largely driven by the farmer. He is closest to the soil and is concerned about groundwater contamination, soil erosion, profitability and passing on a heritage to future generations. To be truly "sustainable," agriculture must produce good food on a long-term basis without depleting the soil or polluting the environment. My criteria for sustainability are zero soil loss, nutrient balance, and 25 earthworms per cubic foot of soil.

There is confusion about biological farming and whether it really works. There seems to be very little research or support from university soil and crops scientists. Will yields decline? Will weeds take over my farm? Will I need to take more time for certain operations? These are good questions to ask as you consider biological farming.

First, let's look at what biological farming is not. It is *not* magic in a gallon jug. It is *not* strictly organic. It is *not* a replay of grandpa's old ways. It is *not* just putting in a few grass waterways to reduce erosion.

Instead, biological agriculture is a *system*, a process with rules and goals. It requires an understanding of what you are trying to accomplish. It works *with* natural laws and beneficial soil life — earthworms, bacteria and fungi. These worms and soil microorganisms change the soil into a loose, crumbly, biologically active soil which resists erosion and soaks up water like a sponge. Soil organisms help release crop nutrients from the soil. This makes for healthier plants which provide their own protection against insect and disease attack.

Weeds are not an inevitable curse but are an index of the character of the soil. Correct the soil's structure and its nutrient balance and you will reduce and alter weed populations; for example, higher calcium and sulfur levels in the soil reduce foxtail.

This is a small taste of the principles behind biological farming and of what you can accomplish in your fields.

So now that you see the "what" of biological farming (namely, understanding your objectives and what you are working with), let's look at some rules for successful biological farming.

Rules for Biological Farming

Farms that are having success with biological farming systems have followed a few basic rules, either by accident or by decision. Here they are:

Rule 1: Test and Balance Your Soil

Good soil is not just a mass of minerals; it is a living thing, with minerals, water, air, organic matter and the organisms that turn organic matter into humus. They are all necessary to grow healthy, high quality crops. But let's emphasize the nutrients needed by plants.

Plants need more than N, P and K (nitrogen, phosphorus and potassium). They need at least 16 elements that have been identified so far. Sure, they need N, P and K, but what about calcium, sulfur, magnesium, zinc, copper, manganese, iron and boron? These, plus a few others, are all needed — in the proper amounts. Some are major nutrients, needed in large amounts, while others are micronutrients or trace elements, of which the plant needs only a tiny amount. These trace elements are of no less importance; they affect health and production in the plant. A *balance* of nutrients is required.

You need to know whether your soil is short of any nutrients, because any nutrient in short supply becomes a limiting factor, reducing yield and quality, and possibly triggering weeds or pest and disease attack. You need to get a good soil test so you know the condition of your soil — in what it is deficient or what it has in too great a quantity. Then begin with first things first and start moving the soil toward a balance. This often means calcium. We also need to provide the crop with a balanced fertilizer that contains the required nutrients that the soil cannot adequately supply.

I like to test for 10 nutrients: phosphorus, potassium, calcium, magnesium, sulfur, zinc, manganese, iron, copper and boron. They are the most important to balance. We do not test for nitrogen but do provide it for the crop. Less nitrogen is needed if the soil is in balance (especially for calcium and sulfur) and if there is a lot of biological activity (soil life). If you are going to put back into the soil what the crop takes off, put on *all* of the essential elements, not just N, P and K.

To properly balance soil, we look at the excesses as well as the deficiencies. We recommend natural, mined fertilizers and small amounts of the highest quality manufactured fertilizers (certified organic farms may not be able to use all of them). We also use kelp, liquid and dry humates, and fish, which provide many essential trace elements, and we try to stimulate soil life and plant roots.

Soil testing is a tool, a guideline to help decide how best to spend your fertilizer dollar. But soils are not mathematical. Putting on 200 pounds per acre of a nutrient will not increase your soil report by 200 pounds. On the other hand, improving the balance of important nutrients and promoting soil life *will* significantly improve your soil's condition and your crops. Adjusting calcium levels is a good place to start. Calcium has a major impact on releasing other elements in the soil, improving soil structure and stimulating beneficial soil organisms. It is absolutely essential for healthy, high quality crops.

So again, Rule 1: Test and balance all essential nutrients. You don't need to test every acre every year. Soils did not get out of balance in one year, and it will take time to correct them. If you have livestock and grow legumes, I suggest testing the fields you plan to seed down. Then put on the necessary fertilizers to bring them into balance. Retest in three or four years as your rotation brings you back to reseeding those fields, to check your progress. More frequent testing may be useful to correct or monitor problem areas.

For crop farmers or those who have not completed soil audits recently, test the *whole* farm. One sample can represent up to ten acres. This will give you an overview of the level of your mineral availability. Retest after a few years of soil corrections and crops to monitor the progress.

Rule 2: Use Fertilizers Which Are Life-Promoting and Non-Harmful

Not all fertilizer materials are the same, and not all soils are the same. Some materials work better under certain conditions, and some contain substances that can harm the soil, plants or soil organisms. Sometimes the cheapest source of an element is not the best source. I prefer materials that are mined with minimal processing, including the following:

1. Idaho phosphate. A natural marine shell deposit containing calcium (30 percent), phosphorus (28 percent), carbon and many trace elements. It provides a continuous (slow release) supply of phosphorus, an element needed for a good root system, normal plant growth and high quality food. Idaho phosphate also works well when mixed with livestock manures. It forms ammonium phosphates, saving the nitrogen content of the manure and making the phosphorus more available. Manure odors are also reduced.

2. North Carolina reactive rock phosphate (33 percent phosphorus, 30 percent calcium). An excellent natural source of phosphorus, calcium and some trace elements. It provides phosphorus more readily than other rock phosphates yet contains slow-release phosphorus for future crop use.

3. Monoammonium phosphate (MAP, 11-52-0). This is a low pH, soluble source of phosphorus, useful for high pH soils.

4. High-calcium lime (calcium carbonate). Local quarry limes can vary. Some are high-calcium lime, but others are dolomite (calcium magnesium carbonate). Dolomitic lime would not be required on high-magnesium soils, plus it is slow to become available. High-calcium lime is recommended on low pH soils (pH 6.0 or under); it is a slow-release type of calcium.

One excellent high-calcium source is Bio-Cal (manufactured for Midwestern Bio-Ag). It has a high solubility and also contains sulfur and boron. It is usually applied at 500 to 1,000 pounds/acre for readily available calcium. Our research shows up to 150 pounds soluble calcium per ton, compared to quarried limes with 5 pounds or less soluble calcium.

5. Gypsum (calcium sulfate). A calcium and sulfur source which works well under conditions where calcium saturation is high and the pH is over 7.0.

6. Potassium sulfate (0-0-50). A high quality mined potassium source which provides both potassium and sulfur. Compared to potassium chloride, it has a low salt index. It is readily available, and its low chloride content does not harm plants or soil organisms.

7. Ammonium sulfate (21-0-0-24S). An excellent source of both nitrogen and sulfur. In the upper Midwest, it works well to reduce high magnesium levels and to provide the sulfur needed for the plant to make high quality protein. Sulfur is also needed to build soil humus.

8. Ammonium nitrate (28-0-0). The source of nitrogen I recommend as a liquid in side-dressing corn.

9. Calcium nitrate. A good nitrogen and calcium source for foliar feeding; also a specialty fertilizer for vegetables.

10. Trace elements. Trace elements in the sulfate or chelated forms are the most effective. They are also acceptable for the certified organic farmer.

Unacceptable Fertilizer Materials

Certain commonly used fertilizer materials may be a cheaper per-unit source of a nutrient, but they are not cheap in the long run because they can degrade soil structure, harm soil life or injure plants. They are too soluble, so they leach easily or cause an imbalance in soil nutrients or a high uptake by plants. A fertilizer element such as potassium or calcium does not occur alone but is tied to another element or ion which is carried along with it. Thus, we could have potassium chloride or potassium sulfate, or we could have calcium carbonate or calcium magnesium carbonate. Some materials I find *not* acceptable, either because of their form, the concentration or the carrier they have, are:

1. Dolomitic lime (calcium magnesium carbonate) is a calcium and magnesium source, sometimes useful on low-magnesium soils. But in some areas, soils are already high in magnesium. Adding more magnesium to the soil does not contribute to balancing it and can interfere with the uptake of other elements, especially potassium. Too high magnesium makes some soils more compact and tight.

Dolomite supplies calcium and magnesium in a 2:1 ratio, two parts calcium to one part magnesium. Ideally, a balanced soil is at a 6:1 or 7:1 ratio. Crops such as alfalfa remove calcium and magnesium in a 5:1 ratio, so with continuous use of dolomitic lime, magnesium levels will get higher and higher, while soils might become tighter. Under these conditions, it takes more of other elements to grow the same crop. Also, tight soils often lead to reduced nitrogen-fixing nodules and poor root health. A lower proportion of calcium also harms crop quality and health.

2. Potassium chloride (0-0-60 or 0-0-62). The potassium is fine, but the high percentage of chloride (about 47 percent) is not. The commonly recommended high doses make the problems worse. Potassium chloride is a strong salt, having the highest salt index of any commonly used fertilizer. It can cause salt damage to seedlings and the roots of sensitive crops and can kill beneficial soil life. Stressed or damaged roots invite root pests and diseases, and crop growth and yield are hurt. It is highly soluble and plants take up large amounts, causing nutrient imbalances and livestock health problems.

Plants only need a small amount of chlorine to grow, which they can get from manures or dust in the air, so adding sometimes hundreds of pounds per acre is unnecessary.

3. Anhydrous ammonia. It may be the cheapest per-unit source of nitrogen, but it will cost you in the long run. Ammonia is a highly toxic gas. It will kill any life near the injection point, including crop roots. Handling and application are expensive. Some ammonia can escape into the air, wasting money. Worst of all, it causes the soil's humus to dissolve and leach, robbing roots of potential nutrients and eventually making soil as hard as concrete. The high concentration of nitrogen upsets the soil carbon-nitrogen ratio, causing a burn-up of carbon in the soil. High ammonia use also causes soil to become acidic.

4. Certain dry fertilizers are not recommended because they can release considerable amounts of ammonia in the soil, harming seedlings or roots. These are diammonium phosphate (DAP, 18-46-0) and urea. Urea in liquid fertilizers is not harmful at normal rates.

5. Oxide-form trace elements. They are cheaper but don't have the availability of sulfates and chelates.

Rule 3: Use Pesticides and Herbicides in Minimum Amounts and Only when Absolutely Necessary

Pesticides and herbicides are made to kill living things. They do not distinguish between good and bad. Just as a crop can be injured by herbicide carry-over, beneficial insects and soil organisms can be killed, thus crippling a natural system for growing good crops. The use of toxic chemicals should be restricted or eliminated.

You will find that the need for pesticides and herbicides will decrease as your soil comes into balance and is more biologically active. Many of the bacteria and fungi in "living" soil protect plants from pests and diseases, and healthy plants are "immune" to pest and disease attack in several ways. When plants have a healthy root system and are vigorously making sugar by photosynthesis, they will seldom have insect problems. A crop's sugar content can easily be measured in the field with an instrument called a refractometer (see Chapter 8).

If you have weed or pest problems, try non-toxic control methods first, such as crop rotation, soil balancing and mechanical control. Rotary hoeing and timely cultivation can control weeds. Some pests can be controlled by releasing natural enemies such as predators, parasites or pathogens.

If non-toxic methods fail — they may take a couple of weeks to control pests — and if crop damage is above the economic threshold, toxic chemicals might be considered. Effective control can often be obtained by reducing the amount of chemical used. Banding a herbicide, spot-spraying, adding humic acid to the tank — make sure the pH of the tank mix is low — using contact herbicides, and proper timing of chemical use are just some ways of reducing amounts of toxic chemicals. Synthetic chemicals can be called "necessary evils" and should be used with that thought in mind.

Rule 4: Use a Short Rotation

When crops are rotated every year or two, there are fewer weed, disease and pest problems. Thus, less herbicides and pesticides are needed and better use is made of green manures and animal manures. Crop yields are also higher and inputs lower than with a long rotation or with a monocropping system.

An example would be one or two years of corn, then beans, then a seeding year, followed by two years of alfalfa. Or, for the grain farmer, a corn/bean rotation interseeded with green manures works well. Just keeping continuous corn or row crops (corn/beans) without a grass or legume green manure crop will keep you from receiving many of the benefits of biological farming.

Rule 5: Use Tillage to Control Decay of Organic Materials and to Control Soil, Air and Water

Good soil should have adequate air and moisture because roots and beneficial soil organisms need oxygen and water. Raw organic matter (plant residues and animal manures) should be tilled into the upper layers of the soil for optimum decay into humus. Leaving it on the surface does little good and may waste nutrients. The humus that is produced will then help improve soil structure, producing better drainage and aeration. An alternate way to add humus is to compost organic matter, then add the compost to the soil. Composting greatly reduces the volume of material to haul.

When soil is tilled deeply, it should not be inverted (turned over), but can be sliced or uplifted. Never till soil that is wet, and keep field traffic to a minimum to reduce compaction. A hardpan can be temporarily broken up by subsoiling, but the best long-term solution is high calcium and adequate humus and soil organisms, especially earthworms and large, deep root systems.

Rule 6: Feed Soil Life

Beneficial soil organisms are a "voluntary army" willing to work tirelessly for you, if only you will let them. If you provide them with a comfortable "home" (soil with air and moisture), food (organic matter), a good mineral balance and freedom from toxic chemicals, they will go to work. If you feed the soil microbes, they will feed the crop. That's how it should work.

Soil organisms do best on a mixture of cellulose-containing (plant matter) and nitrogen-containing (animal manures and legumes) organic matter. About two parts cellulose-containing to one part nitrogen-containing material is a good mixture. The organic matter can be worked into the soil's upper layers or can be separately composted in piles or windrows. Adding rock phosphate or a little lime to compost piles will help produce a more balanced fertilizer. Incorporating a green manure crop, like rye, red clover, Austrian field peas, alfalfa or buckwheat, is another way to feed soil life and improve soil structure.

If raw organic matter is added directly to the soil, allow enough time for it to break down before planting a crop, or else some nutrients will be temporarily "tied up" by the microorganisms, starving the crop. Do not overload the soil with heavy applications of manure or other raw organic matter. It is better to apply a lighter coat over more acres. Applying composted organic matter can be done using any amount, since compost is basically humus already.

Besides animal manures and green manures, biological stimulants may be useful. Many can help improve soil structure and stimulate soil organisms. These products include kelp (seaweed), humic acids, enzymes, vitamins and hormones. Others are inoculants of beneficial bacteria, fungi or algae. Such products often produce results when used in conjunction with good biological farming practices, but you should still get the major nutrient elements in balance before relying on these "fine-tuning" products.

Summary

Now that you have a better understanding of what biological farming is, the "how-to" part should be a little easier. First, know your objectives, your goals. One goal should be to work *with* nature, not to exploit it. You will need to think more, run more field tests, dig in the soil, check root systems, and evaluate alternative practices. It's not a job that can be done from a tractor cab with headphones on.

How long will it take to change over from chemical-intensive farming to biological farming? Remember, your soil didn't get in its present condition in a year. It probably will take several years to get an abused soil into really good shape. During the transition period, you will notice improvement every year. You shouldn't have to sacrifice yields or profits. With proper planning, yields and profits should rise. Don't forget, you're not going to be buying a lot of expensive weed and pest control chemicals. You are just going to be spending your dollar differently. After a few years, your inputs should lower significantly. The advice of a knowledgeable consultant can be valuable during the transition period.

If you are still not convinced that there is a better way to farm, you need to prove it to yourself. Visit a successful farm that is already using biological methods. Observe and ask questions. Don't expect to copy exactly what another farm is doing, however. Usually you will have to pick and choose from several options and adapt and develop a total farming system for your own farm and your circumstances and budget. Again, an experienced consultant can help.

For those who are just starting on biological farming, you are not alone out there. Many more are seeking, searching and experimenting. They are excited about what they are accomplishing.

So, start with information and understanding. Go to soil testing for essential nutrients. Balance your soil with good fertilizers. Reevaluate your tillage and chemical usage. Start feeding the life in the soil, and stop treating your soil like "dirt." You are then on your way to a more profitable, safe, clean and satisfying way of farming.

3

NATURE'S WAY: THE SOIL-PLANT SYSTEM

To better understand what is going on in biological agriculture, we should take a look at the total natural system that grows crops, the *soil-plant system*. You could compare the natural system to a man-made urban society, with factories that manufacture products, raw materials suppliers to the factories, waste-removal workers, and a transportation network to carry things from one place to another. There also might be criminals and murderers who cause chaos, and police who control them. They all work in a structural system of buildings, roads and bridges that have to be built and maintained.

Nature's system. The soil and plants in your fields make up a natural production system, along with the atmosphere and the precipitation that falls. It is a food-producing system. Your crop plants are the food factories. The raw materials are (1) air (specifically carbon dioxide and nitrogen gases), (2) water, (3) over a dozen mineral elements, and (4) sunlight. The sun provides the energy that runs the factory. Through the complex, almost magical, biochemical process called *photosynthesis*, plants combine carbon, hydrogen and oxygen to produce food in the form of sugar. Later, some of the sugar may be transformed into other forms of food, such as starches, oils or fats, proteins and vitamins. Nitrogen and sometimes sulfur are added to sugar to make proteins. Food can be thought of as "trapped" energy from the sun. It is stored energy that can be used by plants for their own growth or reproduction, or by animals and humans, or by microorganisms.

In the natural soil-plant system, raw materials, waste products and finished products move from place to place. They are *recycled*, and the system is a self-sustaining, renewable system. In natural situations with little or no human interference, such as a forest or prairie, the system tends to stay in a dynamic balance with the nutrients and energy continuously being recycled. It can be diagrammed thus:

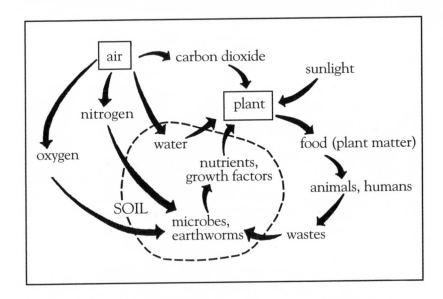

One of the key components of the soil-plant system is soil life — the dozens of kinds of microorganisms, worms, mites and other critters that inhabit the soil. They will be covered in more detail in later chapters. Briefly, they perform a wide variety of tasks. Nitrogen-fixers trap atmospheric nitrogen and make it available to crops. Bacteria and fungi break down complex materials (organic matter) into simpler "raw materials" (plant nutrients) and the humus that improves soil structure. Root microorganisms help feed nutrients to crop roots. Some eat the others, helping to keep their populations in check. Some act as police, arresting or killing the "bad guys" that cause diseases or are parasites. And creatures such as earthworms and ants actually "plow" the soil, mixing and aerating it.

Agricultural systems. Natural soil-plant systems are usually in balance and self-sustaining, but when man comes along and "tinkers" with them, the balance can be upset, and sometimes the system can be completely destroyed. Environmental pollution, habitat destruction and species extinction are well-known examples. Agricultural practices are based on plant-soil systems that are to some degree artificial or interfered with. Livestock grazing on native pasture or woodland management are probably as close to nature as you can get. The more frequent situation of a crop planted in a tilled field is quite unnatural, similar to a city with only one kind of factory (one crop species) and only a few raw materials suppliers and waste haulers (soil life). Perhaps there are too few bridges and police, so supplies have trouble getting where they are needed and criminals are taking over (crop pests and diseases). And when the farmer harvests part or most of the plants, some nutrients are removed from the system.

We could say that whenever a part of the natural soil-plant system is eliminated (such as earthworms killed by pesticides or plant matter harvested), the farmer has to take over the job of the part eliminated if he still wants to grow high-yielding, healthy crops. If nutrients contained in crops are removed and not recycled into the soil, the farmer has to replace them with expensive purchased fertilizers. If earthworms are not there to mix and aerate the soil, the farmer has to till it. If natural pest and disease control agents are crippled, the farmer has to use expensive and dangerously toxic chemical controls. In field crop production, some of this kind of intervention by the farmer is unavoidable, since part of the crop is harvested and weeds have to be controlled. But why cripple the natural system unnecessarily?

The biological farmer will want to do all he can to promote and not harm the many parts of the soil-plant system so that they can do most of the work in growing good crops and maintaining the sustainability of the system. In this way, the farmer can do less of the work and reduce purchased outside inputs to a minimum. That is why I strongly recommend that you learn all you can about the soil-plant system (its parts and how they work) so you can more intelligently plan and make management decisions.

Roots and the system. As an example of how important a knowledge of the workings of the natural soil-plant system is, let's take a look at crop roots and how they interact with the other parts of the system. In general, the bigger and healthier a plant's root system, the better the growth and yield of the top of the plant, and the better the crop can take up nutrients and withstand drought. Roots absorb water and available nutrients from soil particles and the water between soil particles. Roots also release acidic substances that make nutrients in the soil available that were previously unavailable. But roots also interact with soil microorganisms. Special fungi called mycorrhizae live in roots and aid them in absorbing extra water and certain nutrients. Microorganisms that either live in the soil or on the roots trap (fix) nitrogen — that's *free* fertilizer. Other microbes work at extracting unavailable nutrients from soil minerals and organic matter. Some species release growth-stimulating substances for roots, such as hormones and vitamins. Soil microbes that live on the root surface can inhibit or kill disease pathogens and nematodes, so they contribute to root health. What does the root do in turn? Roots release large amounts of carbohydrates into the soil — a kind of "food" for the beneficial microorganisms that live near the root.

It's a beautiful system. When things work right, the soil and all its life forms help feed the plant and keep it healthy, and the plant feeds its helpers. When the root dies, its organic matter also feeds the microbes, and they recycle the nutrients for the next crop.

But the beneficial organisms in the soil need certain things to survive and do their jobs. Almost all of them need:
• well-aerated soil (good soil structure),
• a moderate amount of moisture (not too wet or too dry),
• livable temperatures (warm),
• a source of food (usually some kind of organic matter), and
• freedom from harmful soil conditions (toxic substances, strong salts, wrong pH).

For the most part, they like to be left alone. Excessive tillage usually disturbs their homes (microhabitat) and depletes organic matter. But if soil is very compact or if there is a hardpan that causes too-wet conditions, remedial tillage to at least temporarily correct the problem can "prime the pump" and help get a better population of beneficial organisms going. Wet and poorly aerated soils tend to favor the disease-causing microbes and nematodes.

What Affects Yield?

The yield of a crop depends on efficient crop growth, since what the farmer is interested in is plant matter of some kind — roots, stalks, leaves, or seeds and fruit. All plants (each species and variety) have a built-in genetic potential to grow plant matter. But in order for the plant to produce as much as its genetic potential allows, the environmental conditions in which it grows should be optimal. If any environmental factor is not ideal, it becomes a limiting factor in crop production. Some limiting factors are beyond the farmer's control, while others can easily be adjusted. That is part of biological farming — good management to control as many limiting factors as possible. Let's look at what affects crop yield as part of the soil-plant system.

Genetics/Variety

A plant's genes control its structure (size, shape) and its growth and reproduction (including root uptake, photosynthesis, food translocation and storage, and flowering and seed production). Plant breeders can "tailor make" a crop variety to grow almost any way they want, but usually they select especially for yield (bulk, quantity), disease/pest resistance, and desirable qualities for large-scale production, harvesting, transport and consumer appeal. Plant breeders also select plants which do best under our high-salt, highly soluble, imbalanced fertilizers, and plants that can tolerate the insecticides and herbicides the best. As for corn, it can now be planted a month or more before it comes up and still survive.

What if you don't farm this way? Are these genetics the ones that fit the good biological farmer? As with biotechnology, in many cases we are putting a genetic Band-Aid on a soil balance and health problem. It's at the farmer's expense. Genetically altered plants are always more expensive, and if you eliminate the problem by working with the soils, you won't be spending money for a problem you don't have.

In plant breeding programs, nutritional quality usually takes a backseat — if it is considered at all. If you grow crops with an eye to producing high quality food, try to select the best variety possible for that purpose.

Still, a plant's genetic potential is modified by many environmental factors. Some may increase the yield beyond the usual average, closer toward the genetic potential: such things as optimal nutrient supply all season, stand density or row width, and leaf orientation play a role. Other factors reduce yield, such as weeds, pests and diseases. Many environmental factors interact; for example, temperature and light both affect photosynthesis. I like selecting plants that are vigorous, grow large roots, are "workhorses," and can compete. You have to have the soil and soil nutrients in such condition that allows the crop to express its maximum genetic potential.

Weather

You cannot do much about the weather, but the condition of your soil can modify the effects of weather on your crops. For example, a hardpan may cause root suffocation from wet soil after a rain. Many factors make up the weather:

Temperature

Plants grow faster in warmer temperatures, but within limits. Each species has a lower limit below which no growth occurs, and an upper limit of temperature above which respiratory needs for energy outstrip food energy produced by photosynthesis. For corn these temperature limits are about 50° and 113° F, with optimal growth between 89° and 95° F. There are "cool-season" crops such as barley and "warm-season" crops such as cotton.

Crop scientists use growing degree-days or "heat units" as a rough measure of the accumulating amount of warmth sufficient to cause plant growth. You may see degree-day figures for each week published in a newspaper or extension publication, but remember that those are average figures, and the amount of heat your actual crops receive depends on your local topography and soil conditions. Bare soil between the rows gets very hot, while a cover crop or mulch keeps it cooler. Biological farmers may have an advantage in cool weather, since biologically active soil tends to be warmer due to microbially produced heat (as anyone who has done composting knows).

Light

Plants of course need sunshine to carry on photosynthesis and produce food. Actually, some plants can receive too much light for their genetic capabilities, and high ultraviolet light is harmful (thus the concern over the atmospheric ozone level, since ozone blocks ultraviolet light). Plants also carry on transpiration (loss of water through their leaves) faster in bright light than dim; this could be important when the soil is short of water. The amount of light in a day (versus the length of night) regulates the time of flowering of some plants. Soybean and rice are "short-day" crops, wheat and red clover are "long-day" crops and corn is neither. Also, the amount and intensity of sunshine affect the temperature of plants and the soil (see above).

Precipitation

Precipitation — usually rain or snow — provides the soil and crop roots with needed water, but it can also add a certain amount of nutrients: nitrogen (especially during a thunderstorm) and minerals (washed out of a dusty atmosphere). A decade or two ago, crops received considerable sulfur from air pollution. For optimal growth and high yield, crops need the right amount of soil water throughout the growing season; too much can suffocate roots and too little lowers growth and photosynthesis. The seasonal distribution of precipitation (as characterized by an area's climate) is very important in determining which crops can grow in an area without irrigation. A climate's wet-dry cycles and the amount of water from snow are significant, since the recharge of groundwater is a major source of a crop's water. Storms can be very important, not only for adding nitrogen, but also if they are violent and cause crop damage or delay of harvest. You need to have the soils "weatherproofed." The moisture that comes down needs to soak in and be stored, so that in dry weather it can move up by capillary action and get to the plant. Soil structure, soil life, organic matter, and large root systems assure you success even under adverse weather conditions, which we know at times we will get.

Wind

Wind is important because it increases evaporation of water, both from the soil and from the leaves of plants (transpiration). In either case a good supply of soil water, as would be the case in humus-rich soil, could help a crop through a drought period. Increased transpiration does increase root uptake of nutrients, however. Strong wind can also cause serious soil erosion; again, humus-rich soil with good structure resists erosion.

Soil Structure

Soil structure refers to the physical conditions of the soil. An ideal structure for crop growth has the tiny soil particles clumped together into larger, stable "crumbs," or aggregates. This provides good aeration (for roots and beneficial soil organisms), rapid water intake and good drainage, moisture-holding ability, and easy root growth, plus erosion is reduced greatly. Soil structure is affected by the soil type and particle sizes (sand, silt, clay), organic matter, soil life, crop root systems, tillage, compaction, weather and soil chemistry (pH, nutrient balance, salinity). In general, higher organic matter, higher calcium levels, higher biological activity, large root systems, minimal tillage, little compaction, and low-salt fertilizers are some management practices that foster good soil structure.

Soil Chemistry

The soil is an extremely complex mixture of various chemical substances — some useful, some harmful and some neutral.

The pH of the soil is a measure of its acidity or alkalinity, technically the hydrogen ion content of the soil's water. The main effects of pH on crop growth are that it affects the availability of many nutrients and it affects the activity of soil microorganisms. The best overall range of pH for most crops and soils is from pH 6.2 to 6.8. But one should realize that the actual pH can fluctuate greatly from day to day and according to the distance from roots, so it is not necessary to be overly concerned as long as the pH on a soil test is close to the "ideal" range (some soil types normally have a more acid or a more alkaline pH than the ideal).

The other main influence of soil chemistry on crop growth is from the type and balance (proportion) of nutrients and other substances in the soil. In order to be *available* for absorption by crop roots, nutrients must either be dissolved in the soil's water (in soil solution) or be loosely held on soil colloids (very small particles — clay and humus). Any nutrients that are tightly held on colloids or are part of the molecular structure of soil minerals or undecayed organic matter will be *unavailable* to roots. In most soils, the great majority of nutrients are unavailable at any one time. The immediately available nutrient supply may not be sufficient during peak demand by crops (one or more nutrients in short supply may be a limiting factor). But some of the unavailable nutrients can gradually become available throughout the growing season, by such actions as weathering of minerals, root-produced acids, root uptake (allowing additional nutrients to go into solution), and microbial activity (microbes can "break out" tied-up nutrients, and some root-inhabiting microbes actually "feed" nutrients to the roots). Certain chelating substances found in humus and released by roots and soil microbes, will make unavailable nutrients available to roots. The biological farmer has many advantages in improving a crop's nutrient supply (and keeping outside inputs low) because he will try to encourage beneficial soil life, increase soil humus levels, use slow-release fertilizers, and possibly use chelating materials (chelated trace elements, humic acids, kelp). Keeping soil calcium levels high and working toward a good balance of nutrients also helps increase nutrient availability and biological activity — thus giving better crop growth and higher yield. "Free nitrogen" can be obtained from nitrogen-fixing microbes, both on legume roots and away from roots in biologically active soil.

Problems in crop nutrient availability can occur especially on non-biological farms. Tight, compacted soil (poor aeration and poor drainage) can restrict root growth and activity as well as microbial decay of organic matter causing less nutrient release. An imbalance of nutrients can slow crop growth or result in low crop food value (low quality). High salts can slow growth or injure roots. Toxic elements, such as aluminum, chlorine or heavy metals, as well as toxic herbicides or pesticides, can slow growth or kill plants. Even applying excess manure or other raw organic matter can stunt growth from toxicity. Good soil structure, high soil humus, beneficial soil organisms and use of non-harmful fertilizers and control agents can alleviate or eliminate these yield-limiting problems.

Soil Biology

Earlier sections have covered many of the positive effects of beneficial soil organisms on soil structure, soil chemistry and crop growth. Since soil conditions greatly affect the well-being of these helpful organisms, the biological farmer will always consider what his actions and inputs will do to (or for) his soil life.

Planting

An obvious factor that affects crop yield is when and how crops are planted. A too-early planting date can lead to frost damage, but if that doesn't happen, then early-planted crops often have an advantage in growth and yield (especially if there is summer drought). The main factor in early planting is soil temperature, not calendar date. High biological activity in the soil warms it up. Small grains will germinate in the 30s Fahrenheit, while corn needs the 40s. Germination and root establishment occurs faster with higher temperature, so later-planted crops usually catch up with early-planted ones; also cold, wet soils encourage root and seedling diseases. Many weed seeds tend to germinate in cool soil, giving them a head start on crops. Some biological farmers delay planting until after early weeds have been cultivated. Any condition less than ideal at planting is unacceptable. Why start out to fail?

A good seedbed is important for good germination and stands. No matter what tillage and/or planting method is used, be sure soil structure and aeration are good, and that soil moisture is not too high or too low (plant deeper in dry topsoil). Preplanting tillage can bring weed seeds to the surface, creating a potential weed problem. Of course, planting depth, row width and seed spacing affect the stand and population. A high population *can* mean a higher yield, but be sure the soil's water and nutrient supply is large enough to feed the crop (irrigation would eliminate that concern). Narrow rows or high plant populations can also cause yield reductions if there isn't enough light (cloudy weather) or if high humidity encourages fungal diseases; however, dense stands of crops are good for shading out weeds.

Weeds

Every farmer knows that weeds can significantly limit yields. Weeds compete with crop plants for soil water and nutrients, and for light. But weeds do not grow indiscriminately everywhere. Different species tend to grow better in certain soil conditions (although some will grow in a wide range of conditions). Most of the weeds infesting fields grow best in poor or out-of-balance soil conditions, while crops do best in good soil conditions. For example, bindweeds and crabgrass grow well in crusted, hard soil with low humus; while foxtail, fall panicum, quackgrass, velvetleaf and buttonweed prefer wet, tight, poorly aerated soils which are low in oxygen. Many farms with these weeds have exactly those types of degraded soils. A biologically active soil with good aeration, high humus and high calcium usually does not have a problem with such weeds (see Chapter 19 for more details). However, a few weeds that are less of a problem do grow in good soil, including lambsquarters, redroot pigweed and common milkweed.

Biological farming methods reduce the worst weed problems because soil structure and balance are improved and because soil organisms are increased. Some microbes improve crop growth, giving crop plants an advantage over many weeds (shading them out). Some noxious weeds grow best in out-of-balance soil, such as low calcium (dandelion, dock, burdock, Johnson grass, red sorrel), or low phosphorus and high potassium (velvetleaf).

The high amounts of soluble fertilizers and/or manures applied early in the growing season set up conditions for the growth and fertilization of weeds. The crop planted can't use all those nutrients that early, and high-solubility fertilizers slow soil life and decrease crop root growth, making crop plants less competitive.

That's why in a fertilizer we want to balance soluble components to slow-release components so that some nutrients are available later. The decay cycle is also very important. Mixing old residues and green growing manure crops shallowly in a soil speeds the decay so that nutrients are released during the mid-season, when crop needs for nutrients are greatest and the crop is able to compete with small, undernourished weeds. Fertilizer balance, placement, concentration and recovery are all a part of growing crops biologically and reducing fertilizer and chemical inputs. Many farmers notice their weed species changing after they begin using biological methods.

Pests

Such crop pests as insects and nematodes (or occasionally others, such as mites and slugs) can certainly cause great damage to a crop, generally by eating plant tissue, or by allowing disease pathogens to invade, or by injecting toxic substances into the plant. Many factors can affect pest attack, including the crop variety (some are resistant), weather conditions, crop vigor and soil conditions. Often pests will only attack those plants growing in a low spot of a field, and upon investigation you may find the soil there to be waterlogged. That is a clue to a general principle of biological agriculture: pests and diseases usually only attack plants that are under stress of some kind. It may be stress from cold, cloudy weather (low photosynthesis), wet soil (low soil oxygen, roots suffocate), out-of-balance nutrients (deficiency of an element needed for proper growth), or perhaps stress from the farmer's use of toxic fertilizers or herbicides (see Chapter 20 for more details).

A vigorous, healthy plant will seldom have pest problems because healthy plants produce various substances that repel or even kill pests or their larvae. Using insecticides (or nematicides, etc.) to kill pests often leads to worse pest problems in the future, partly because beneficial enemies of pests such as ladybugs, lacewings, and many others also are killed. Pesticides may cause pests to become resistant to the poisons used against them or cause the plant's vigor to be decreased. It is even possible that the "quick fix" provided by the pesticide obscures the farmer's concern about the condition of his soil.

Biological farmers always put their soil's condition and the encouragement of beneficial forms of life high on their list of priorities. Healthy plants just don't have the same level of pest problems. Many farmers find that their need for insecticides drops to zero after a year or two of using biological methods. Some biological methods that help in decreasing pest problems include a good crop rotation which discourages soil-inhabiting pests, interplanting (growing different species of plants close together repels some pests or confuses their plant-locating ability), and keeping stress-causing factors as low as possible (by practices such as maintaining good soil structure and a proper nutrient balance).

Diseases

There are a large number of plant diseases, and they can be caused by various pathogens, including viruses, bacteria and fungi. Nematodes are responsible for certain diseases also. These invading pathogens either destroy plant tissue or release substances that impair normal plant functions.

But just as we noted above for pests, it is a general rule that diseases are a much greater problem in stressed plants. That is the real cause of diseases and pests. Vigorous plants have natural disease-fighting mechanisms, similar to the way our bodies fight off most diseases until stress lowers our immunity. Many things can cause crop stress — bad weather, poor soil conditions, use of toxic fertilizers or control chemicals, and so on. Planting a resistant variety, if possible, and maintaining biologically active soil with good structure will usually keep diseases to a minimum. Many of the beneficial soil microbes actually inhibit or kill pathogens. Some do this by producing antibiotics, such as penicillin and streptomycin. Again, a good crop rotation will often reduce soil-borne diseases significantly.

Crop Root System

A very important key to high yields, which is often overlooked, is a large and healthy root system — of course, some crops are roots, such as carrots and beets. Roots normally provide the crop's supply of water and nutrients, although some nutrients can come in through the leaves, as in foliar feeding. Scientific studies have shown that the yield of a crop is correlated with the size of the root system. This is partly because the larger the root system, the more soil the roots can explore for water and nutrients. Drought resistance is better. A larger root system also anchors the plant better and withstands whatever root pests or diseases that may attack. A healthier, more vigorous root system is more active in absorbing water and nutrients. Roots have a hard time growing in dense, compact soil, but earthworm burrows stimulate root growth. Healthy roots have a high population of beneficial microbes — bacteria and fungi — on their surface which help them in absorbing water and nutrients, and even protect them from pests and disease pathogens.

Biological farmers will think about their crop's root system, and occasionally dig up root systems to check on their growth and health. They will maintain soil with good structure and high biological activity.

Practical Hints

As a summary of this section, let's list some practical things you can do to increase crop yield and quality:

- Plant a variety that can give high yield (genetic potential, resistance to pests and/or diseases), that fits your soil types and farming methods, and that produces quality crops.
- Plant a high enough population (be sure soil water and nutrients are sufficient). As the soil improves, you can increase the plant population.
- Eliminate physical soil restrictions such as compaction, hardpan or drainage problems.
- Provide timely applications of balanced fertility (see Chapters 11 and 12).
- Recycle organic matter (animal manure, green manures, previous crop roots and residues), and foster soil life. Don't just rely on outside fertilizer inputs.
- Do everything possible involving soil structure to weatherproof a crop. Practices to achieve this may include subsoiling, zone tillage, promot-

ing biological activity (use green manure crops to feed soil life), and root stimulation. If crop stress develops from weather or insufficient fertility, try foliar feeding.

Our yield potentials are much greater than what we have been producing, as the winners of crop-growing contests regularly demonstrate. Over-200-bushel corn and over-60-bushel soybeans, with less input than conventional systems use, are easily possible with today's knowledge and biological methods.

What Affects a Crop?

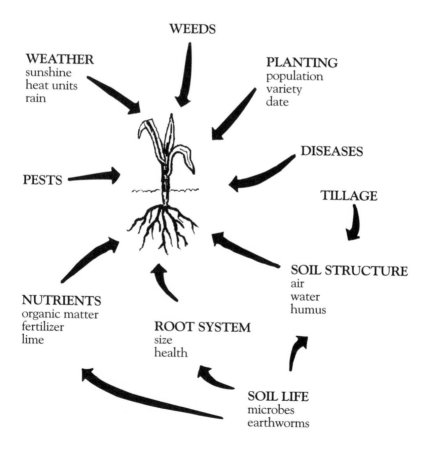

WEEDS

WEATHER
sunshine
heat units
rain

PLANTING
population
variety
date

DISEASES

TILLAGE

PESTS

SOIL STRUCTURE
air
water
humus

NUTRIENTS
organic matter
fertilizer
lime

ROOT SYSTEM
size
health

SOIL LIFE
microbes
earthworms

4

SOIL BASICS: WHAT IS IT?

Soil is the basis for agriculture. Our food crops are grown in soil, and our livestock and poultry eat the crops. We either eat plant or animal food, and we also grow fiber crops and make many industrial and medical products from plants. Our soil has been called our most important resource. Wise use and management of our soils are vital for maintaining modern society. Our entire civilization depends on a thin layer of topsoil, often only inches deep. Mistreatment of the soil is the number one source of crop problems — and later, animal problems. Through misuse, about seven to 10 tons of topsoil per acre are being lost through erosion each year in the American Midwest. It may take as much as several hundred years for one inch of soil to form. In addition, once-fertile soil is being polluted by toxic pesticides and herbicides or is being made into wasteland by salt accumulation. We can't go on treating our soil like "dirt."

Soil Parts

What is soil? It's not simple. Soil is a complex mixture of several components, capable of supporting plant life. It originally formed by the weathering of the earth's rocks. Typical soil contains approximately the following proportions of four main parts:

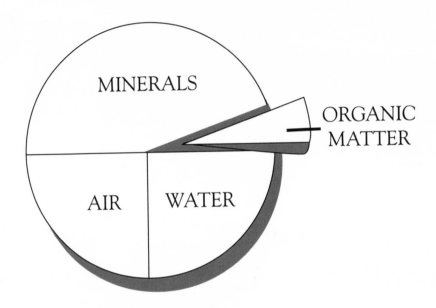

1. Minerals — about 45 percent
2. Water — about 25 percent
3. Air — about 25 percent
4. Organic matter & living organisms — about 1 to 5 percent

    The soil's minerals come from the underlying rocks. Some of the mineral matter supplies needed plant nutrients such as calcium, magnesium, phosphorus and potassium, while part is insoluble or not used by plants and contains aluminum, silica and iron oxide. Most of the mineral nutrients in typical soil are "locked up" in the molecules of the mineral particles so they are not available at one time to plants. But a small part of these mineral nutrients can become available by the action of weathering, root-produced and microbial-produced acids, and chelating substances. Note the total mineral and nutrient content of typical soils in the following table:

Composition of Typical Soils

(for a plow layer 6⅔ inches in depth, approximately 2,000,000 pounds)

COMPONENTS	SANDY LOAM (lbs/acre)	SILT LOAM (lbs/acre)	CLAY LOAM (lbs/acre)
Organic matter	20,000	54,000	96,000
Living portion; microbes, earthworms, etc.	1,000	3,600	4,000
Nitrogen	1,340	3,618	6,432
Silicon dioxide	1,905,000	1,570,000	1,440,000
Aluminum oxide	22,600	190,000	240,000
Iron oxide	17,000	60,000	80,000
Calcium oxide	5,400	6,800	26,000
Magnesium oxide	4,000	10,400	17,000
Phosphate	400	5,200	10,000
Potash	2,600	35,000	40,000
Sulfur trioxide	600	8,500	6,000
Manganese	2,500	2,000	2,000
Zinc	100	220	320
Copper	120	60	60
Molybdenum	40	40	40
Boron	90	130	130
Chlorine	50	200	200

(Compiled by J.L. Halbeisen & W.R. Franklin.)

As you can see from this table, there's more to soil than N-P-K and lime. Plants need over a dozen mineral elements. The total amounts listed in the table are not all available to plants at one time. But this is the biological farmer's secret. Through soil structure changes, large root systems and biological activity, the farmer can help nature release some of these tied-up minerals. Each soil is different, and farming methods differ. The soil system can provide a certain amount of nutrients, sometimes all for that season. In other cases, some nutrients need to be added in the proper balance and from proper sources.

Soil Particles

Typical soils contain particles of varying size. Based on size, the minerals of soil can be classified (with their relative surface area) as follows:

| | SIZE | | SURFACE AREA | | |
| | diameter | | of 1 cubic centimeter or 0.06 cu. in. | | |
PARTICLE	millimeters	inches	sq. cm	sq. in.	sq. feet
Coarse sand	2.0-0.2	0.08-0.008	60	9.3	0.065
Fine sand	0.2-0.02	0.008-0.0008	600	93	0.65
Silt	0.02-0.002	0.0008-0.00008	6,000	930	6.5
Clay	< 0.002	< 0.00008	60,000	9,300	65
Colloidal clay	< 0.0005	< 0.00002	600,000	93,000	650

Most soils contain at least some sand, silt and clay-sized particles. You can see them and their relative amounts by thoroughly shaking up a jar filled with half soil and half water. Let the jar sit overnight. Larger particles settle out first, then the smaller ones. The smallest, colloidal-sized particles won't settle out, but remain suspended in water (making it look cloudy).

Soils are named according to the relative proportions of the different particles; for example, a sandy soil, a clay soil, a silt loam or a sandy clay loam. A loam has roughly equal parts of sand, silt and clay.

Colloidal particles (colloids) are among the most important parts of soil, for it is on them that exchangeable crop nutrients are held for possible use by plants. Note in the above table the very large amount of surface area that colloids have, compared to sand- or silt-sized particles. Both clay and humus can contain colloidal particles. Their large surface area is one reason they can hold a lot of nutrients. The other reason is that their surface is covered by electrical charges, like static electricity. Most of the charges are negative, meaning they can attract and hold positively charged nutrients. In fact, humus colloids can hold about three times as many nutrients as clay (see Chapter 5).

Soil Pores

In typical, good soil, solid particles make up only about 50 percent of volume. The other 50 percent is pore space (actually, it can be from about 30 percent to 75 percent in various soils). The pore space can be filled with either air or water. In a well-drained soil of moderate moisture content, about half the pore space will contain water and half air, but either the air or water portion could be as little as ⅓ or as much as ⅔ of the pore space in average soils, depending on the soil structure and the amount of soil moisture. Obviously, whatever pore space is not filled with water will be filled with air (not considering roots or soil organisms).

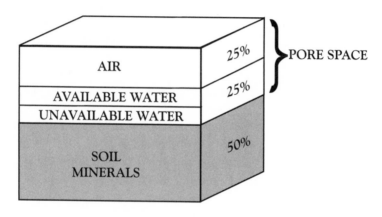

Soil Air

Oxygen is essential for root growth, water and nutrient uptake, and for biological activity (soil life). Most of the beneficial forms of soil organisms require oxygen. Good soil needs to be well aerated. Soil with plenty of air or oxygen is said to be *aerobic*. Poorly aerated soil with little or no oxygen is called *anaerobic*. When soil is anaerobic (for example in waterlogged or very dense soil), roots can "suffocate" and stop growing and absorbing, and most of the beneficial organisms either die or become dormant. Certain bacteria that can live without oxygen (anaerobic species) can cause problems by releasing nitrogen into the air (denitrification) and producing toxic by-products such as hydrogen sulfide, methane, ammonia and others. Certain soil elements, including ferrous iron, sulfide, cobalt, molybdenum, boron, zinc, nickel and lead, can become too available and build up to toxic levels in anaerobic soil. All these stresses may later lead to crop diseases and pest attack.

Roots and organisms in the soil respire. They use up oxygen and give off carbon dioxide. Some carbon dioxide dissolves in soil water, some escapes into the air through soil pores which plants can then use as a raw material in photosynthesis. A well-aerated soil will allow easy gas exchange between the atmosphere and soil life. Additionally, well-aerated soils warm up faster in the spring.

Soil Structure

Good soil structure is essential for successful farming. As we have seen, soil air is vital for healthy roots and beneficial soil life. The ideal soil in most types of agriculture has a loose texture, with its tiny mineral particles glued together into larger clumps, called aggregates or crumbs, much like the texture of cake. These glued-together aggregates resist erosion. A handful of soil should crumble easily. There will be about 50 percent pore space, and about half of that will be filled with air and half with water. A commonly used term for soil structure is *tilth*, referring to ease of tilling. Soil with good tilth should not have a surface crust or a subsurface hardpan; it should be loose throughout.

Compaction is an enemy of good tilth. It squeezes out air and destroys good structure. Compaction can result from the weight of farm machinery, but soil with poor structure and low organic matter will tend to become naturally compact as its particles move closer together. A compact subsurface layer, or hardpan, can be formed either from tillage tools or naturally from soil particles moving closer together, often as a result of water percolation. Soils with an imbalance of certain elements can also become compact through chemical or electrical attraction of particles.

Soil Water

The water in soils is necessary for plants, since plant tissue contains about 80 to 90 percent water. But water is especially important because it dissolves and carries nutrients and other materials inside the plant and in the soil. Soil water always has materials dissolved in it, so it is often called the *soil solution.*

If a soil is thoroughly saturated by water and then allowed to drain (under the effect of gravity), that water which drains is called *gravitational water,* and the drained soil is said to be at *field capacity.*

Because some of the pores in soil are very small and because electrically charged colloidal particles attract water molecules, not all of the soil water is usable by plants. Typically about half the water at field capacity is available water, with the rest being unavailable to plants. When the available water is gone, the soil has reached its wilting point, which is the point at which a plant begins to wilt and cannot recover.

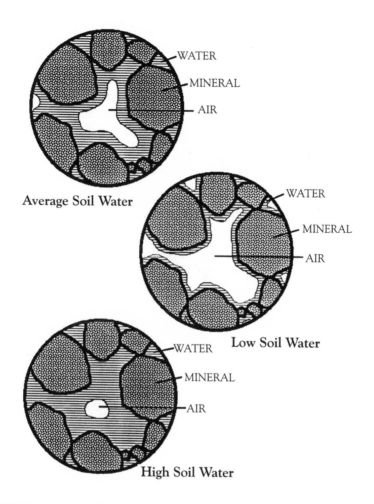

Average Soil Water

Low Soil Water

High Soil Water

SOIL TYPE	PORE SPACE(%)	FIELD CAPACITY(%)	AVAILABLE WATER(%)
Sand	30	10	7
Sandy loam	40	18	12
Loam	50	26	17
Silt loam	54	36	22
Clay loam	62	40	20
Clay	65	45	18
(Compiled from various sources.)			

The amount of water a soil can hold at field capacity — and therefore the amount available to plants — depends on the soil type. In sandy soils, the total pore space is small, but much of the water can be used by plants. In clay soils, there is a lot more pore space, but many pores are so small and so much water is tightly held that plants can't get as large a percentage; however because of the larger pore space, clay soils do hold much more plant-available water than sands.

Soil Organic Matter

Although it makes up less than 5 percent of volume in most soils, the soil's organic matter has an importance far out of proportion to its amount. F.E. Allison, in his book, *Soil Organic Matter and Its Role in Crop Production* (1973), calls organic matter "the key to soil fertility and productivity." Dr. William A. Albrecht has called it "the constitution of the soil." It plays major roles in the chemical, biological and physical aspects of soil fertility.

Organic matter is not just one material; it is a complex, dynamic mixture of substances. We can divide it into two categories:

1. Fresh ("raw") and incompletely decomposed plant and animal residues, typically such things as animal manures, crop residues and old roots, and green manures. These materials still retain some of the characteristics of the original materials from which they came, such as plant stems or root remains. In moist conditions, they will probably be undergoing decay by various microbes (bacteria, actinomycetes, fungi) and possibly earthworms.

2. Humus. This is decomposed organic matter which is dark brown or black in color and has virtually no remains of stems, roots, etc. Humus is a complex mixture of the chemical building blocks of life (carbohydrates, fats, amino acids, amides, tannins, lignin, etc.) plus by-products of the microbes that broke them down (organic acids, aldehydes, alcohols, ketones, enzymes, vitamins, polysaccharides, etc.). The polysaccharides (complex carbohydrates) play a role in gluing small soil particles together and creating good soil structure. Humus is rich in certain large molecules called humic substances, including humic acids, fulvic acids and ulmic acids. They aid in making nutrients available to plants, improving soil structure and in stimulating plant growth.

Both the raw and decomposed organic matter are important in soil. The raw materials do aid in giving bulk to soil and reducing erosion, but their main function is as food for soil microbes and earthworms. The decomposed organic matter, *humus*, has many benefits. Humus:

a. Is a storehouse of crop nutrients: 90 to 95 percent of the soil's nitrogen, 15 to 80 percent of phosphorus, 50 to 70 percent of sulfur, and some potassium and trace elements.

b. Increases the available supply of nutrients, releasing them from "tied-up" forms. Humus also chelates nutrients (holds them in an available form) which would otherwise become tied up.

c. Increases the soil's cation exchange capacity (CEC) or ability to store nutrients. Humus contains colloidal particles (see above, "Soil Particles") which can attract and hold nutrient ions. In typical soils, humus furnishes 30 to 70 percent of the soil's nutrient-holding capacity.

d. Increases the soil's water absorbing and water holding capacity. Humus acts like a sponge, and its particles hold water so roots can absorb it.

e. Contributes to good soil structure by gluing together soil mineral particles into larger "chunks" called aggregates or crumbs. This produces larger pore spaces in the soil and thus improves aeration and water movement (drainage), as well as making the soil less dense, which improves root growth. In diagrammatic view, we can see how organic matter glues together soil particles.

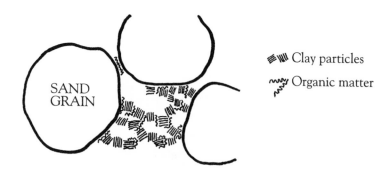

SAND GRAIN

▤▥ Clay particles

〜〜 Organic matter

f. Acts as a buffer in the soil. A buffer resists changes in pH or ion level. Thus, humus protects plants from high salt levels, toxic ions, and drastic changes in pH (acidity or alkalinity).

g. Moderates soil temperatures, helping to warm soil in cool weather by increased microbial activity, aeration and darker soil color (absorbs more sunlight).

h. Stimulates beneficial soil organisms by providing a food source and growth stimulating factors (vitamins, enzymes).

i. Stimulates plant growth and health, not only by the above benefits, but also by providing growth stimulating factors (hormones, vitamins, enzymes).

As measured by most soil testing labs, "organic matter" is actually a measure of the carbon content of the soil, which usually corresponds to the organic matter level. The "percent OM" on your test report would include raw undecomposed material, decomposed humus, and any of the living organisms that were also in the soil sample. The most important part is humus, so some testing labs will also run a "humus test" (actually a humic acid test).

The approximate amount of organic matter in soils can be estimated by their color and soil type:

SOIL TYPE	AVERAGE % ORGANIC MATTER
Sands & sandy loams	0.3-1.8
Dark sandy loams & light silt loams	1.8-3.3
Dark silt loams & loams	3.3-4.2
Dark clay loams	4.2-5.5
Mucks & peats	over 20

A virgin soil, such as in a forest or a prairie, has a certain stable level of organic matter, depending on the vegetation and climate. Agricultural activity, especially tillage and crop removal, and possibly use of certain fertilizers and toxic chemicals, will gradually deplete the natural soil organic matter, reducing it to a new lower level. Typically, about 40 percent of the original level will be lost in 15 to 20 years of farming. Tillage causes more rapid oxidation and microbial decay. Row crops and the erosion that accompanies them are very detrimental. Good crop rotations and soil management, such as growing sods and cover crops and recycling animal and green manures, can maintain agricultural soils' organic matter at a higher level than under poor management. That is a major concern of biological farming.

Biological Farming Objectives

To maintain good soil conditions . . .

1. Always be aware of soil structure, humus and beneficial soil life and their importance.

2. Avoid practices that can destroy good soil structure:

a. Tillage when soil is too wet.

b. Unnecessary tillage.

c. Tillage tools that can compact soil or cause a hardpan.

d. Unnecessary field traffic or driving on wet soil.

e. Use of fertilizers that can destroy humus or harm soil life, such as high rates of ammonia, dry urea, DAP (diammonium phosphate), or high-salt fertilizers (potassium chloride, or muriate of potash or kalium potash).

f. Use of herbicides or pesticides that can kill beneficial soil microbes and/or earthworms.

3. Keep a relatively high proportion of calcium and sulfur in the soil. These elements improve soil structure and stimulate beneficial soil life. Correct any imbalance of soil elements.

4. Frequently recycle organic matter. Add a small to moderate amount of fresh organic material and work it into the upper layers of the soil, add compost, or grow crops with fine root systems (especially grasses). This feeds humus-producing organisms as well as directly improving soil structure. The element sulfur is necessary to build humus, and organic matter supplies some sulfur.

5. Manage your nitrogen applications. Use only as much as is absolutely necessary. Place it when and where it is needed. Excess nitrogen reduces soil biological activity and interferes with carbon-nitrogen ratios. The excess nitrogen "burns" or uses up the carbon to get back in balance, and that is burning up organic matter and humus.

6. If soil has a chronic compaction or hardpan problem, use deep tillage to break up and aerate the soil. This should "prime the pump" and allow beneficial organisms to proliferate. If they are fed with organic matter, they should eventually build up humus and reduce the compaction.

7. Reduce erosion:

a. Do not farm steep hillsides, or at least do not plant row crops.

b. Plant rows along the contour, not up and down a slope. Use contour strips or terraces on steeper slopes.

c. Keep the soil covered. Grow ground covers and grass water ways, and interplanted crops in row crops. Leave a mulch of crop residues on the surface of bare soil.

d. Build soil humus levels.

e. Promote soil life, especially earthworms.

8. Provide an abundant supply of soil minerals in the proper balance.

9. Make sure water soaks in where it falls.

10. Make sure water is stored and maintained until plants need it. Get it into the soil, protect it from evaporation, and recover it for crop use.

a. Eliminate barriers to water intake from crusting, compaction or a hardpan.
b. Build a high humus level. Humus soaks up and holds water like a sponge. It's built-in drought-proofing.
c. Maintain good soil structure. Soil pores allow capillary recovery of deeper water ("wicking" of water upward through narrow pores).

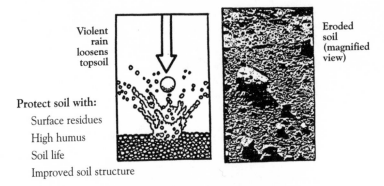

Violent
rain
loosens
topsoil

Eroded
soil
(magnified
view)

Protect soil with:
  Surface residues
  High humus
  Soil life
  Improved soil structure

d. Reduce water loss by surface mulching, residues or a cover crop. In dry weather, shallow cultivation breaks a crust and keeps soil from excessive drying by providing a "dust mulch."
e. Promote soil life. Beneficial organisms increase humus and improve soil structure. Earthworm burrows increase water intake. A type of beneficial fungus (mycorrhizae) lives in plant roots and helps them absorb water.

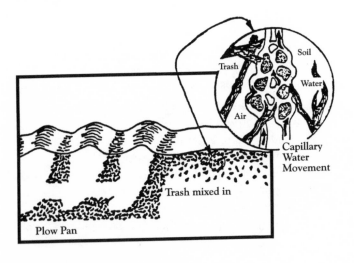

Soil

Trash

Water

Air

Capillary
Water
Movement

Trash mixed in

Plow Pan

5

SOIL BASICS: HOW IT WORKS

The soil works as a dynamic plant-growing system. Soil, along with the atmosphere, precipitation and sunshine, provides all the needed elements and conditions for optimal crop growth — or at least it should.

Crop Needs

Plants need at least sixteen elements to carry on life processes, to grow and reproduce. Most of a plant's bulk comes from the air, not the soil.

TYPICAL PLANT COMPOSITION (%)

Oxygen	45.0	
Carbon	44.0	95% FROM AIR
Hydrogen	6.0	
Nitrogen	2.0	
Potassium	1.1	
Calcium	0.6	
Sulfur	0.5	
Phosphorus	0.4	

TYPICAL PLANT COMPOSITION (%) (cont.)

Magnesium	0.3	**5% FROM SOIL:**
Manganese	0.05	• Minerals • Humus
Iron	0.02	• Fertilizers
Zinc	0.01	
Chlorine	0.01	
Boron	0.005	
Copper	0.001	
Molybdenum	0.0001	

(Compiled from various sources.)

Actually, recent research has shown that most plants also require very small amounts of nickel, and some plants require silicon.

Note from the table that plants need varying percentages of different elements. In other words, they require a certain balance of nutrients from the soil, more nitrogen, quite a bit of potassium, but more sulfur and calcium than phosphorus. And they need *all* elements, not just N, P and K. They need trace elements also, in very small amounts. And plants need varying amounts of each element throughout the growing season — a little at first, then much more, then less as they finish maturing. How does the soil supply all these nutrients — in the right balance and at the right time?

First, it should be pointed out that most soils do not completely meet all of the plants' needs all of the time. That is partly why crops seldom, if ever, fulfill their genetic potential (adverse conditions other than the soil also lower yields, such as bad weather). But *if* the soil were ideal, how does it nurture and sustain growing plants? There are many ways.

Water Uptake

Plants need water as part of their cells. Plant tissue contains about 80 to 90 percent water. The water in cells is a substance in which the various molecules and ions that perform the metabolic activities are dissolved or carried (included are such activities as photosynthesis, food transport, cell maintenance and growth, and reproduction). But water is also the substance in which the soil's nutrients are dissolved and carried into the plant.

Plants are constantly losing water, so a continuous supply must be provided through root absorption from the soil. Most water loss occurs through a plant's leaves; it is called *transpiration*. Leaves have thousands of tiny pores (called stomata or stomates) that allow water to evaporate out (but they also let carbon dioxide in for use in photosynthesis). A single corn plant might lose over two quarts of water per day, or over 50 gallons in its lifetime. Plants can regulate the size of their stomata and can slow down water loss in dry periods. Another way water is lost is in the root exudates that "leak" out of the tips of roots (to be covered later).

This constant loss of water through the leaves does have a purpose. The upward flow of water carries nutrients from the soil to all parts of the plant (in fact, the loss of water from the leaves exerts an upward pull which causes most of the absorption by roots at the bottom of the "pipeline"). Another function of transpiration is to cool leaves in hot weather.

Down in the soil, water fills part of the pore space, the area between soil particles. In moist soil, about half of the total soil water is able to be absorbed by roots (called available water) and about half is too tightly held on soil particles to be used (unavailable water). Soils with a high clay content hold a higher percentage of the total water as unavailable water, but their total water supply will be larger than in sandy soils. Another factor affecting root water uptake is the salt content of the soil. High salt levels make it harder for roots to absorb water (a plant can wilt from lack of water in salty soil, even though the soil has plenty of water). Some soils are naturally salty, as in the western United States, but excessive use of high-salt fertilizers (such as potassium chloride or ammonium nitrate) can also increase salt levels so high that crop growth is seriously harmed. Some crops can tolerate more soil salt than others. Biological farmers avoid overuse of salt fertilizers.

Roots usually have thousands of tiny root hairs near their tips which extend out into the soil. Root hairs give the root a much larger surface area (five to 20 times larger) and allow more efficient absorption from a larger volume of soil.

Soil water near the root tip enters root hairs and the epidermis cells that cover the root. Then water flows through and between the cells of the outer zone (the cortex) and into the root's center where the "plumbing" (vascular tissue) is. There are two types of vascular tissue, *xylem* which carries water and nutrients upward, and *phloem* which carries food from the leaves downward to other parts of the plant. The xylem cells are like long pipes, and they fit together end-to-end to form a tubular water-carrying system that runs from root tips through the stem and into the leaves through the veins of the leaf.

The faster the rate of transpiration from the leaves, the faster the roots absorb more water. Transpiration and water uptake occur faster at higher temperatures, during the day, during sunny weather, in low humidity

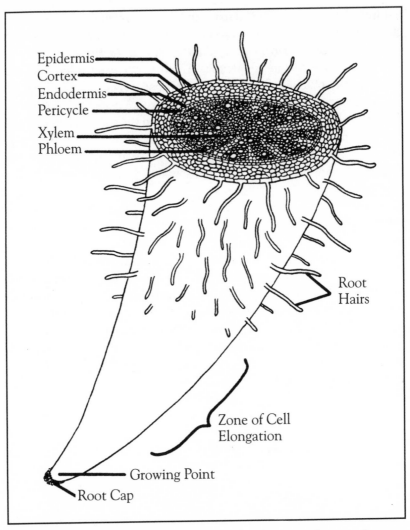

Epidermis
Cortex
Endodermis
Pericycle
Xylem
Phloem

Root Hairs

Zone of Cell Elongation

Growing Point
Root Cap

weather, and in windy weather. That is why plants need so much water during hot, dry spells. If the soil's water supply is low, or if the soil is "tight," with very small pores that restrict water flow, a plant that is rapidly transpiring from its leaves will not be able to absorb enough water through its roots, and it will begin to wilt. It may be able to recover during the night, when transpiration is slower. This is one reason the soil's structure and humus content are so important: to improve water movement and availability.

A large, healthy, actively growing root system is important for providing "drought resistance" to a crop. Roots grow best in loose, uncompacted soil (good soil structure), especially with a lot of earthworm burrows (which allow easy growth and are lined with nutrient-rich castings). Roots also grow best in soil with a good balance of nutrients, including sufficient calcium, phosphorus, nitrogen and boron. Biological farmers try to consider what effect their operations and additives will have on crop root systems.

Crop nutrients. As we mentioned at the beginning of this chapter, plants need over a dozen nutrient elements, which they mainly get from the soil (we will ignore foliar, or leaf uptake for now). Plants need various amounts of nutrients in differing proportions throughout the growing season. For example, here is a soybean crop's (60 bu/acre) uptake of five of the dozen soil nutrients (in lbs/acre):

NUTRIENT	40 days	80 days	100 days	120 days	140 days	TOTAL
Nitrogen	9.1	140.4	10.8	74.4	73.2	307.9
Phosphate (P_2O_5)	1.3	24.0	3.6	14.4	14.4	57.7
Potash (K_2O)	7.3	118.8	8.4	45.6	44.4	224.5
Calcium	2.9	34.8	8.4	13.2	0	59.3
Magnesium	0.7	10.8	1.2	6.0	3.6	22.3

(Compiled from various sources.)

Note from the table that nutrient needs are small at the beginning of the crop's growth, then increase dramatically during peak vegetative growth and during seed production. That's why management of the decay cycle and providing a balance of nutrients (both soluble and slow-release) are so important for producing healthy, high-yielding crops. Management of the decay cycle means that the residues are digested by mid-summer. Then and only then can their nutrients be released. This also is the time the plant has the largest need for nutrients.

We saw in Chapter 4 that soils have far larger amounts of nutrient elements than crops need, but most of this total soil nutrient supply is unavailable to plants. Most of the nutrients in soil are "tied up," with their molecules chemically bound in mineral particles, or in the complex organic molecules in humus or the bodies of soil organisms. Some of these unavailable nutrients can become available through natural processes: (1) weathering, the action of precipitation and temperature changes; (2) root release of acidic substances (hydrogen ions, organic acids); and (3) microbe release of acids and chelating substances and microbial decay of organic matter. In a healthy, biologically active soil, these natural release mechanisms can often meet much of a plant's nutrient needs (although in soils of low cation exchange capacity or with high plant populations per acre, this nutrient supply can fall far short of crop needs, so supplemental fertilizers are necessary). Still, natural nutrient release is one reason biological farmers can reduce their fertilizer inputs after several years.

Roughly 5 percent of a typical soil's total nutrient supply can become available each year and held on soil particles or in soil water. Nutrients that are held (by chemical or electrical bonds) on soil particles are called *exchangeable nutrients*. Plants use nutrient elements in the form of *ions*. Ions are electrically charged atoms or molecules. They have either a positive electrical charge (called *cations*) or a negative charge (called *anions*). The nutrient ions used by plants (those considered essential for plant growth) are:

ELEMENT	SYMBOL	IONS USED BY PLANTS
Nitrogen	N	NO_3^- (nitrate), NH_4^+ (ammonium)
Phosphorus	P	$H_2PO_4^-$ (orthophosphate, dihydrogen phosphate), HPO_4^{--} (hydrogen phosphate)
Potassium	K	K^+ (potassium ion)
Calcium	Ca	Ca^{++} (calcium ion)
Magnesium	Mg	Mg^{++} (magnesium ion)
Sulfur	S	SO_4^{--} (sulfate)
Iron	Fe	Fe^{++} (ferrous), Fe^{+++} (ferric)
Copper	Cu	Cu^{++} (cupric)
Zinc	Zn	Zn^{++} (zinc ion)
Manganese	Mn	Mn^{++} (manganous)
Boron	B	BO_3^{--} (borate), $B_4O_7^{--}$ (perborate)
Molybdenum	Mo	MoO_4^{--} (molybdate)
Chlorine	Cl	Cl^- (chloride)
Nickel	Ni	Ni^{++} (nickelous)

(In some of the above formulas, O stands for oxygen and H stands for Hydrogen.)

The very small colloidal particles in the soil, clay and humus particles have a *large* surface area (see Chapter 4, "Soil Particles"). They have many negative electrical charges covering their surface. Positively charged ions (cations) are attracted to and held by these soil colloids (positive and negative charges attract, similar charges repel). Clay particles generally have an angular shape (they are crystals of mineral) and are composed of several flat plates or layers stacked on top of one another, somewhat like a deck of cards.

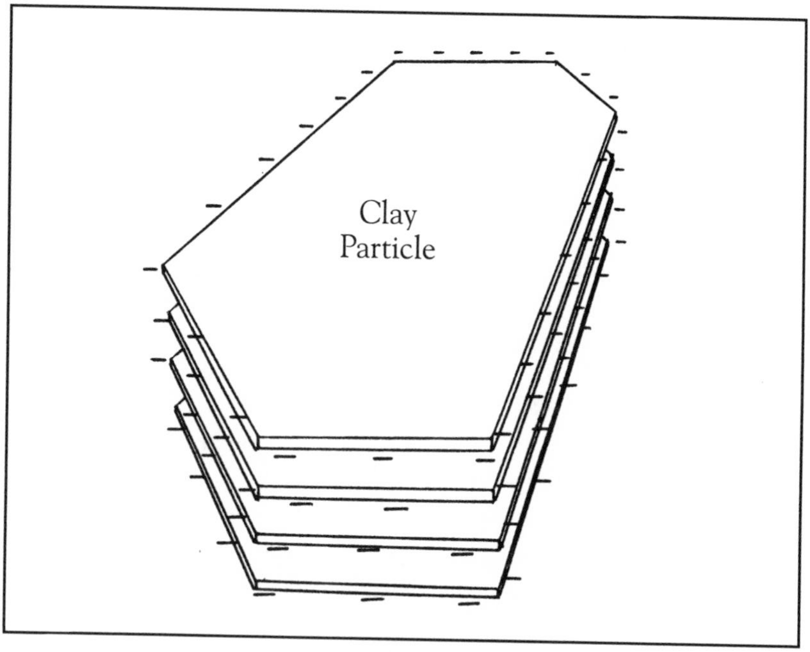

A clay particle with its negative sites "loaded" with cations would look like this (smaller cations can sometimes also fit in between the plates):

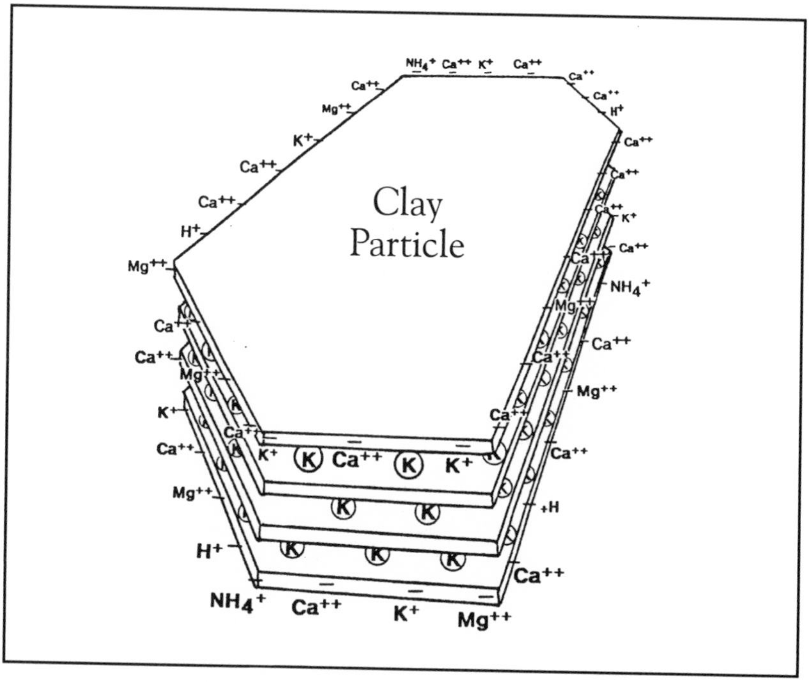

Humus colloidal particles are more irregularly shaped (rather "lumpy"). Humus particles can hold about three times as many nutrients as clay particles.

Negatively charged nutrients (anions) are mainly found free in the soil solution (water between soil particles), although a small amount can be held on soil particles.

Again, the nutrient ions held on soil particles are called *exchangeable*, meaning they can be exchanged from the particles to the root. The soil solution also has some nutrients dissolved in it, usually a smaller amount than what is held on particles. These two nutrient sources together make up the soil's available nutrients, able to be absorbed by roots. However, some of the cation nutrients held on soil colloids are held very tightly, so are sometimes termed *difficultly available* (potassium is an example). Another term for nutrients that are difficult for plants to absorb is *fixed* (tightly held on soil particles).

Soil scientists try to measure how many nutrients a soil can supply to a crop (a kind of "fertility" measure). They calculate what is called the soil's *cation exchange capacity*, or CEC. The CEC measures the soil's ability to hold only the cations (mainly calcium, magnesium and potassium) and not other nutrients, but it is a fairly good indication of a soil's fertility. Obviously, the more clay and humus colloids a soil has, the higher its ability to store nutrients, so the CEC reflects soil type.

SOIL TEXTURE	CATION EXCHANGE CAPACITY
Sands	1-5
Sandy loams	5-10
Loams, silt loams	5-15
Clay loams	15-30
Clays	over 30
Peat	10-30

(Compiled from various sources.)

Humus alone has a CEC of 100 to 300, while pure clays can vary from less than 10 to 200 (different kinds of clay particles vary in their negative sites).

You could think of your soil's CEC as its "nutrient gas tank." It is the amount of "fuel" (nutrients) the soil *could* hold (often translated to yield potential). Some soils have a larger "gas tank" than others:

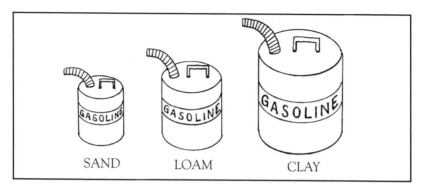

SAND   LOAM   CLAY

But, even if your soil has a large "gas tank," it may be "running on empty" (depleted of some or most nutrients), and your crops could be "out of gas." Poor early growth, diseases and pests, and low yield or poor quality are indications. Plants can be low in one or more nutrients *before* you can see visible deficiency symptoms such as fired or reddish leaves.

In other words, a soil might have a lot of colloidal particles and be able to hold a lot of nutrients, but the available nutrient supply could be depleted, as happens when high yields of crops are grown for several years with little or no fertilization or organic matter recycling. When this happens, the negative sites on the colloids become depleted of nutrient cations (calcium, magnesium, potassium) and become filled up with hydrogen ions (H^+). A sign of this is soil acidity, or low pH (pH is a measure of hydrogen ions, which cause part of the soil's acidity).

An acid or low pH soil is "bad" mainly because it is usually a nutrient-poor soil. Some soil consultants recommend liming acid soil only to "correct" the low pH, but the real reason is to resupply depleted nutrients. The choice of liming material and fertilizer you use should depend on which cations the soil needs for a good balance — calcium, magnesium or potassium (see Chapter 10). The soil's pH can vary from place to place — next to a root it can be as much as two or three whole numbers lower than the surrounding soil — and soil pH also changes after a rain and throughout the growing season. Therefore, it is not wise to be overly concerned by the exact pH numbers on a soil test; they may be plus-or-minus a whole number at different times. Try to take soil samples at the same time each year.

Nutrient uptake. Roots take in nutrients by different mechanisms. Some nutrients — those dissolved in the soil solution — simply flow into the root along with the water it is absorbing. Sometimes nutrient ions also will diffuse (slowly "seep") into roots because the concentration of their ions is greater outside the root than inside. Cells can also "pump" ions inside against the normal flow of diffusion, called *active transport.* Another mechanism is called *base exchange*, or cation exchange (cations are also called *bases*). The root releases hydrogen ions from its tip. The hydrogen ions then "trade places" with cation nutrients held on soil colloids. The freed cations can then easily be absorbed into the root by the above methods. The hydrogen ions fill up the colloid negative sites, and the soil becomes more acid. The "gas tank" runs empty and needs to be refilled.

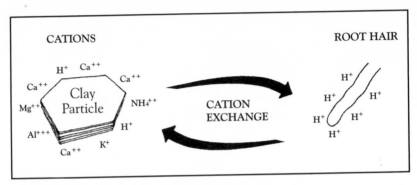

As mentioned earlier, natural processes will release some tied-up nutrients, especially in biologically active soil, but supplemental fertilizers are usually necessary for most soils and for high yields.

The actual amount of a particular nutrient roots take up can be affected by the balance, or proportion of the nutrients in the soil. The positively charged cations are generally absorbed according to their relative abundance, so too much of one will mean too little of others (there is only a certain total amount a soil can hold, measured by the CEC). That is why soil balance is so important to biological farmers, who are interested in crop *quality* as well as quantity (yield). For most soils and for most crops, a good "ideal" balance to try to achieve is about 70 to 85 percent of the total cations being calcium, 12 to 18 percent magnesium and 3 to 5 percent potassium. This measure is called *percent base saturation.*

A soil with an "ideal" balance of these cations could still run short of nutrients part way through the growing season if it is a low CEC soil, such as a sand or sandy loam. It just doesn't have a large enough "gas tank" to hold the needed nutrients. For high yields and healthy crops, the use of balanced, low-salt starter fertilizers should benefit. The result is a concentration of nutrients where the most recovery takes place. Splitting fertilizer applications (especially mid-season nitrogen) and/or foliar feeding throughout the season should be beneficial.

Even if you have a high-CEC soil and a "full gas tank," other factors could interfere with a plant's actually taking up the nutrients it needs. Again, that's the benefit of low-salt balanced fertilizer placed in the row for high recovery. We don't live in an ideal world, and it is impossible to predict the type of growing season you will have. That's why one year test plots may or may not show the benefits of some of these practices. But in the long run, it does pay off.

For optimum crop growth, health, yield and quality, your soil also needs:
• Good aeration.
• Adequate moisture (not too much or too little).
• Sufficient organic matter (2-3 percent or more).
• Proper pH (6.0-6.8).
• Proper balance of *all* nutrients (not just N-P-K).
• High biological activity (microbes, earthworms).
• Plants with large root systems.

The Role of Soil Life

As we have mentioned briefly before, the beneficial organisms that inhabit good fertile soil play an amazing role in helping crops grow. Now let's look at them in more detail.

Because they live down in the soil, the average person is not aware of the large numbers or the many types of soil organisms. There may be two or three *tons* of living organisms in every acre. They are so important that some consultants call them your "soil livestock."

Kinds of Soil Organisms

Most soil organisms are either small or microscopic, so again, they escape our notice easily. The only common large type is the earthworm. Here is a summary of the most common kinds:
• *Bacteria.* Microscopic, one-celled. Most prefer neutral to alkaline soils. Aerobic (oxygen-using) species are beneficial; many anaerobic ones can cause toxicity or disease.
• *Actinomycetes.* Microscopic, thread-like. Need aerobic, neutral or alkaline soils.
• *Fungi, or molds.* Microscopic, thread-like (colonies can be large enough to see). Beneficial species are aerobic and prefer acid soils. The type called mycorrhizae lives in plant roots and aids plant growth.
• *Algae.* Microscopic (colonies can be large enough to see), one-celled. Contain chlorophyll, make food by photosynthesis; near or on soil surface. Help improve soil structure.
• *Protozoa.* Microscopic, one-celled animals. Prefer moist soils; mainly eat bacteria.

- *Nematodes, or roundworms or threadworms.* Tiny, thread-like. Most are harmless; some cause plant damage or disease.
- *Earthworms.* Easily visible, with segmented body. Help recycle organic matter, aerate soil and enrich it with their castings. Improve soil structure and crop growth.
- *Arthropods.* Small to larger, with six to eight or more legs (except some larvae). Includes mites, ants, termites, beetle and fly larvae, spring-tails, millipedes and centipedes. Most are harmless; some cause plant damage or disease. Their burrowing may aerate soil.

An idea of the relative numbers of these soil organisms can be obtained from the following table, giving populations for a temperate grassland soil.

ORGANISM	NUMBER PER ACRE	LBS PER ACRE
Bacteria	800,000,000,000,000,000	2,600
Actinomycetes	20,000,000,000,000,000	1,300
Fungi	200,000,000,000,000	2,600
Algae	4,000,000,000	90
Protozoa	2,000,000,000,000	90
Nematodes	80,000,000	45
Earthworms	40,000	445
Arthropods	8,160,000	830

(Adapted from: L.M. Thompson & F. Troeh, *Soils & Soil Fertility,* 4th ed., 1978, p. 111.)

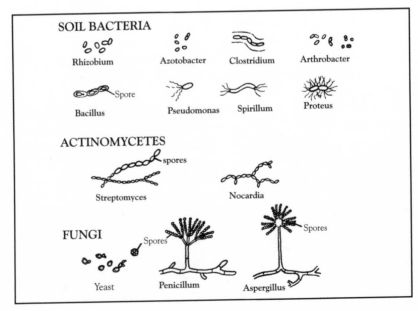

Functions of Soil Life

Soil organisms perform an amazing number of beneficial functions, including:

- Decompose raw organic matter to release nutrients and produce humus.
- Release nutrients in soil mineral particles.
- Fix ("trap") nitrogen from the air.
- Slowly "feed" nutrients and water to roots. Some (mycorrhizae) obtain water and nutrients from farther away in the soil than roots can reach.
- Temporarily "tie up" nutrients, keeping some from leaching and allowing slow release later in the season.
- Produce plant growth stimulants (vitamins, enzymes, hormones).
- Protect roots from invading disease pathogens, nematodes and certain insects.
- Improve soil structure (by making humus and by gluing soil particles together); reduce erosion; increase soil aeration; increase water uptake, retention and drainage.
- Detoxify soil by breaking down toxic chemicals.

Organic Matter Recycling

Soil organisms eat first. Whenever raw organic matter is added to the soil, nearly all of the potential crop nutrients it contains are "tied up" as part of large complex molecules, such as carbohydrates, fats, proteins and nucleic acids. Plants need small simple molecules or ions. Raw organic matter needs to be decomposed by soil microorganisms to release tied-up nutrients for plant absorption. Earthworms and other scavenger organisms in the soil also aid in organic matter recycling, but the majority of the work is done by microbes.

But as they are breaking down the complex organic substances into simpler substances, the microbes use most of the nutrients first, before plants can get them, to build and run their own cells. Only after they die and decompose are most of the nutrients available for plant use. The cells of microbes are approximately equivalent to a 10-5-2 fertilizer (plus they also contain small amounts of the other necessary plant nutrients besides N-P-K). As they decompose organic matter, many beneficial organisms also produce other helpful substances: plant growth stimulating hormones and vitamins, and protective antibiotics that keep pathogens under control.

In practical terms, this means there is a time lag between the application of raw organic matter and crop availability of nutrients. We can't give a definite figure because it depends on soil conditions. Recycling takes place faster in warmer temperatures and at moderate moisture. Good aeration is also important, since in anaerobic conditions (low oxygen), decay is slower and many toxic by-products are produced. In moist, warm soil it might take two weeks or less. This means that you should not add a lot of raw organic matter just before planting (wait at least two weeks) because the microbes will temporarily "rob" the crop of some nutrients, and some of their by-products can injure seedlings. Fall application may work better. Using a starter fertilizer should overcome any nutrient problem, however. Keep raw organic matter in the upper layers of the soil (do bury it, but not too deeply), so it will stay moist but be well-aerated. Composting raw organic matter before putting it in the soil is a way to eliminate the above problems, but the same problems can arise if the composting is not done properly (see Chapter 18).

For best activity and high populations of soil life, give them a good "home":

- Enough food — frequent application of organic matter
- Enough water — keep water mobile within the soil (good soil structure)
- Not too acidic — lime to best pH (usually 6.0-6.8)
- Not too cold (too much surface trash; possibly compaction)
- Not too hot (bare exposed soil surface)
- Watch the toxic materials; they can kill soil life
- Anhydrous ammonia
- Highly soluble salts and excess chloride
- Insecticides, herbicides and other control chemicals

Agriculture is at an infant stage in understanding and working with beneficial organisms. The types of tillage, residues, fertilizers and green crop additions to a soil also affect the numbers and types of organisms. If fibrous, woody materials are added, you end up with a fungus-dominated food web. Some crops like this, but others do poorly. Green crops and low-fiber manures promote soil bacteria. The different organisms and numbers can be altered by farming practices. As more information about the soil food web is obtained, we will know what residues or green manures will benefit which crops. For most crops, a balance between bacteria and fungi is desirable.

Earthworms

Earthworms are a special, larger-sized kind of soil organism which have many beneficial effects on the soil and crops. Consultant and earthworm researcher Dr. Bill Becker, Central Illinois Research Farm, Springfield,

Illinois, estimates that a good earthworm population can save farmers as much as $20,000 to $30,000 per year for an average-sized farm. Earthworms have been called "nature's plowmen" and are recognized by agronomists as important. Many conventional crop production methods are very harmful to earthworms, such as monoculture of row crops, excess use of toxic and high-salt fertilizers, soil compaction and toxic pest and weed control chemicals.

Earthworms prefer to live in rich soil with plenty of organic matter, although a few live in decaying organic matter such as manure. They do best in soil with moderate moisture and high calcium. Excessive disturbance when they are active and near the surface, as from tillage, tends to reduce populations. Sandy soils usually have few or no earthworms because of their typically low organic matter and low calcium levels.

The main benefits of earthworms to the soil are:

- *Improve soil structure.* As earthworms "eat" their way through the soil (or eat plant residues directly), they excrete their wastes as "castings," or casts, either on the surface as little lumps or plastered along the sides of their burrows. Earthworm casts act as soil aggregates, and very "stable" ones resistant to break-up. Well-aggregated soil is aerated and resists erosion. Soil compaction and even a hardpan can often be reduced or eliminated by earthworm activity. They aid in increasing soil humus, which itself improves structure and water retention.
- *Mix and aerate soil.* By their burrowing activities, earthworms churn and mix the soil, and their burrows provide a system of air-filled tubes that allow air to enter, as well as increasing water intake. Surface-applied fertilizers, lime or organic matter can be gradually incorporated by "nature's plowmen," although they seldom can do this at a fast enough rate to be effective for that year's crop.
- *Improve soil pH.* Whether the soil is too acid or too alkaline, earthworm populations tend to move the pH toward the ideal range. They excrete wastes, stimulate soil microorganisms and increase humus, all of which modify and buffer soil pH.
- *Release plant nutrients.* In various ways, earthworms increase the amount of available crop nutrients. They speed up the cycling of organic matter, and increased microbial activity also releases nutrients from organic matter and/or soil minerals. Earthworm casts are higher in available nutrients than ordinary soil. One test of casts found 4.6 times more nitrogen, 1.4 times more calcium, 3 times more magnesium, 7.2 times more phosphorus, and 11 times more potassium than surrounding topsoil. The nutrients in earthworms' bodies become available when they die; they contain about 12 percent nitrogen, so

average annual nitrogen release from dying worms can run 16 to 82 pounds/acre, according to scientific estimates.

By increasing soil humus levels, earthworms can be responsible for raising the soil's CEC (cation exchange capacity), or its nutrient-holding ability. Earthworms' stimulation of beneficial microbes also indirectly aids plant nutrition, since these microorganisms make nutrients available and may "feed" them to roots (see below).

- *Stimulate crop growth.* Besides improved growth from better nutrient availability, plants also can grow better from growth-stimulating hormones and vitamins in the soil. The probable source of these substances is from the soil microorganisms that earthworms increase (humus also stimulates plant growth). Roots especially show better growth when earthworms are present. In various tests, crops sometimes gave double to 10 times higher yields with earthworms than without.
- *Improve crop health.* Through their soil improvement and microbe-stimulating effects, earthworms have an effect on plant health and disease and/or pest resistance. Studies have found 37 to 66 percent reductions in cyst nematodes and reductions in soil-borne fungal diseases.

The biological farmer will always be concerned about the size and well-being of his earthworm population. To build up a good population of earthworms (I aim for at least 25 earthworms per cubic foot of soil in the spring or fall), and to keep them "happy":

- Maintain good soil conditions: balanced fertility, high calcium, moderate moisture, low compaction.
- Cut crop residues into small pieces and incorporate them shallowly. Any aggressive major tillage (chisel, moldboard plow) is damaging, especially in the fall. Deep subsoiling for good drainage may be necessary at some times.
- Feed them raw organic matter, such as plant residues and some animal manures. They like moderate amounts of high-fiber animal manures, such as cattle manure, but can be killed by high applications of strong or liquid manures. These are best fed by incorporating these materials near the surface.
- Maintain a ground cover or mulch of plant residues in the fall for protection from sudden cold spells in the fall or winter. They want their food on top, and they want to be left alone.
- Reduce or eliminate toxic chemicals. Some herbicides and pesticides are not very harmful to earthworms, but others are very lethal. Some of the worst chemicals include chlordane, endrin, heptachlor,

aldicarb, benomyl, carbaryl, phorate, chloroacetamide and pen-tachlorophenol. Tests of anhydrous ammonia in North Dakota found that injecting this toxic gas in September when earthworms were still active reduced populations to one-third of what they were after a November application, when worms had burrowed deep for the winter. The test's control (no ammonia) was at the same level as the September application, showing that earthworm populations do well with a good level of soil nitrogen. However, adding a legume to the crop rotation gave as good a response (reported in Better Crops, vol. 78, no. 3, pp. 10-11).

Root-Inhabiting Soil Life

One of the most amazing and interesting aspects of what goes on down in your soil is the close relationship between plant roots and several kinds of microbes that live close to or even inside roots. The area immediately next to roots is called the *rhizosphere*, and those organisms that live there are known as rhizosphere species. Some of them are hardly found anywhere else in an active state, and they help roots grow and stay healthy in various ways. Except in decaying organic matter, microbial activity in the bulk of the soil is usually at a rather low level. But in the rhizosphere, a width of a few millimeters, the abundance and activity of microorganisms is typically several times, up to 20 or even 2,000 times greater. A major reason is that roots release a watery material called root exudate. This exudate comes mainly from the tips of roots, and it contains food substances for the beneficial microbes, including sugars, amino acids, fatty acids, organic acids, vitamins, enzymes and mucilages. Tests have found that from two to 18 percent of the sugar produced in the leaves by photosynthesis flows out of the roots in exudates (it can be transformed into the other compounds besides sugar). Root exudates can also function by adding moisture to locally dry soil and in releasing nutrients held on soil particles, but exudates have a major importance in stimulating beneficial rhizosphere organisms.

The most abundant rhizosphere organisms are bacteria, with fungi and actinomycetes being less common. There can be 50,000 to 3.25 billion bacteria in a single gram of rhizosphere soil. The main beneficial activities of rhizosphere microbes are:

- *Supply plant nutrients.* Nitrogen-fixing bacteria use atmospheric nitrogen for their own needs, but when they die nitrogen from the air becomes available for plant use (free fertilizer). Some nitrogen-fixing species live inside nodules on crop roots (legumes), but others (non-symbiotic species) are found separately in the soil. A field of alfalfa can fix as much as 200 pounds/acre of nitrogen, while non-symbiotic nitrogen-fixers can add from one to 14 pounds/acre.

Rhizosphere microbes can also release *chelating agents*, substances that increase the availability of many soil nutrients, especially trace elements. Acids and chelating agents can also "break loose" some of the nutrients tightly held in soil mineral particles, such as phosphorus, potassium, magnesium and calcium.

Microbes that decompose organic matter release the nutrients tied up by it, especially much nitrogen, phosphorus and sulfur. The microbes first use the nutrients for their own needs, but they later become available to roots.

A special type of fungus called *mycorrhizae* lives partly inside plant roots, yet does not act as a parasite or disease pathogen (they are symbiotic). The mycorrhizae grow slender cells out into the surrounding soil, reaching farther than the root's root hairs. They can increase the root's absorptive surface area by as much as 100 times. They absorb water and nutrients that the root could not reach, so they improve the crop's nutrition and drought resistance. In return, the root gives the fungus some of its exudate food. Mycorrhizae have been found to increase crop uptake of phosphorus, nitrogen, sulfur, zinc, copper and molybdenum, and in low-phosphorus soils, phosphorus uptake has increased from 30 to 500 percent with corn yields 50 percent higher (unfortunately, there is no yield effect in high-phosphorus soil). Normally, sprouting seedlings will pick up mycorrhizae from the soil, but such factors as fungicide use, flooding (anaerobic conditions) and long fallow periods can seriously reduce the soil's population of mycorrhizae. Then crops can suffer from phosphorus or other deficiencies.

- *Growth promotion.* Rhizosphere microbes release plant growth promoting substances, including vitamins and hormones.
- *Protection.* Several kinds of bacteria, actinomycetes and fungi actually protect roots from pathogens (disease-causing microbes) and plant-attacking nematodes. They have an amazing arsenal of tricks. Some bacteria and fungi crowd out harmful species by multiplying faster on the root's surface — they out-compete them. Many rhizosphere species inhibit or kill pathogens by releasing natural antibiotics. You have heard of some of them because they are manufactured commercially: penicillin, streptomycin, aureomycin, terramycin, chloromycetin and tetracycline.

Certain soil fungi have an unbelievable way of controlling nematodes; they produce special cells that are either sticky or shaped into a "noose." When an unlucky nematode pokes its body through a "noose" or touches a sticky cell, it is immediately trapped and digested by the fungus. Unfortunately, because of conventional agricultural practices and use of fungicides, these beneficial fungi are now rare in U.S. fields.

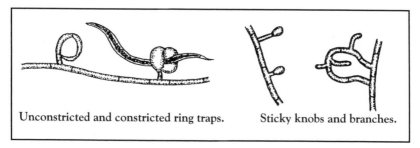

Unconstricted and constricted ring traps.   Sticky knobs and branches.

Some fungi grow onto and digest other, pathogenic fungi (those that cause root rot, for example).

Some of these root-protecting and growth-stimulating microbes are now being used in biological products to improve crop growth and yield, either as seed-treats or soil inoculants (see Chapter 16).

This has been a rather long and involved chapter, showing many of the workings of soil and the interrelationships between crop plants and the soil, and soil life. It's a very complex subject, and you don't have to know all the details to be a biological farmer. Still, the more you understand, the more you will appreciate the amazing soil-plant system and how natural processes, natural systems and nature's beneficial soil organisms work together to help your crops grow better.

The biological farmer will keep these things in the back of his mind, and when he is thinking about tilling a field or spreading manure, he will stop and consider: "What will this activity do to my soil microbes, or my soil structure?"

Remember

Soil life needs a good home. Don't destroy it by improper tillage.

Soil life needs a constant supply of food during the warm part of the year. Light coats of manure, green manure crops and shallowly incorporated crop residues are the best sources.

Air and water need to be controlled. Over-aeration does damage, as does under-aeration, such as waterlogged soils. Observe your soil. Study and look to know your objectives.

Evaluate chemical and highly soluble fertilizer use. Proper management can reduce the use of both.

Finally, soil life needs a proper supply of minerals. A soluble calcium source is essential, and proper carbon-to-nitrogen ratios need to be maintained.

6

SOIL BASICS: BALANCE

In nature, everything tries to be in balance. Predators and parasites keep prey species in check. Pest insects have their natural enemies to control their population levels. Natural laws, processes and cycles keep everything in balance. It is often called the "balance of nature."

Man is only a temporary steward of the soil. Our food — our life — ultimately comes from the soil. Plants use sunshine, carbon dioxide, water and soil nutrients to make food by photosynthesis. Animals and humans eat the plants, and plant and animal wastes are returned to the soil for recycling by microorganisms.

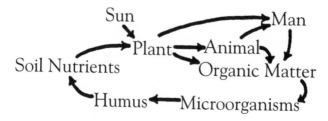

This cycle, with all its organisms and interactions, is often called an *ecosystem*. Because it has so many organisms that rely on each other for their food and survival, an ecosystem is easily upset. Too much or too little of something will cause changes throughout the cycle. The balance will be upset. This often happens when people destroy natural habitats (for example, marshes, prairies or forests) or eliminate species.

Agricultural ecosystems are not completely natural and can easily be thrown out of balance, partly because of the smaller number of species involved (low diversity). Growing only one species of plant in a field allows pests to quickly overrun it if the plants are stressed. An agricultural ecosystem requires frequent monitoring and careful management. Ecological studies have found that if there is greater species diversity, as when crops are interplanted or when there are fence rows or nearby meadows with a variety of plants and insects, there are fewer pest problems in the crops. Interplanted crops make it harder for pests to find their desired plant species, and predators or parasites from fence rows can help control pests.

The ladybird beetle or "ladybug" is a common predator of aphids and other small insects and mites.

Soil Balance

In the soil, balance involves three areas: chemical, physical and biological. All are related to each other.
- Chemical = minerals and nutrient elements used by crops to grow.
- Biological = soil organisms — they decompose raw organic matter to make humus, which holds water and nutrients.
- Physical = the soil's texture and structure, which relates to aeration and drainage. Tillage and soil life affect the soil's physical properties.

All three are tied together and are equally important for production of high quality crops.

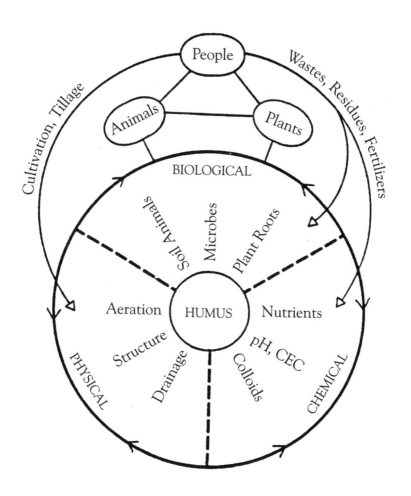

People

Animals   Plants

BIOLOGICAL

Cultivation, Tillage

Wastes, Residues, Fertilizers

Soil Animals   Microbes   Plant Roots

Aeration   HUMUS   Nutrients

Structure   Drainage   pH, CEC   Colloids

PHYSICAL   CHEMICAL

We can visualize soil balance and summarize the signs of balance versus imbalance by the following:

BALANCE: Chemical – Physical – Biological

OUT OF BALANCE	IN BALANCE

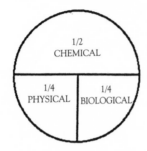

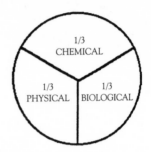

SYMPTOMS:

1. Soil harder, compacted.

2. Soil doesn't dry out, muddy.

3. Small root systems.

4. Crops can't tolerate drought.

5. Plants not healthy; cannot provide health to livestock.

6. Weed and pest problems.

7. More chemicals & fertilizers required.

8. Bigger tractors, more horsepower needed to farm.

9. Less profit.

RESULTS:

1. Soil easier to work.

2. Improved crop quality.

3. More efficient use of fertilizers.

4. More profit (the bottom line): lower inputs, few weed and insect problems.

PHYSICAL BALANCE

OUT OF BALANCE
Common situation

| 75%
SOIL MINERALS | 12 1/2%
WATER |
| | 12 1/2%
AIR |

IN BALANCE
Ideal profit situation

| 50%
SOIL MINERALS | 25%
WATER |
| | 25%
AIR |

SYMPTOMS:

1. Poor water drainage; runoff.

2. Waterlogged, pot-hole problems.

3. Hardpan.

4. Ground cloddy, hard to till.

5. Poor residue decomposition.

6. Weed problems.

RESULTS:

1. Good water intake and retention.

2. Mellow soil; easy to work.

3. No crusting or hardpan.

4. Reduced erosion.

5. Large root systems.

6. Fewer weeds; improved herbicide performance, less required.

CAUSES:

1. Too much of wrong types of fertilizers, humus destroyed.

2. Excessive nitrogen use; nutrient imbalance.

3. Working soils with wrong tools at wrong time.

4. Poor organic matter management (till in too deep, too much).

5. Too many toxic chemicals; soil life is harmed.

REASONS:

1. Use of life-promoting fertilizers to achieve balance.

2. Abundant soil organisms.

3. Proper tillage.

4. High humus content.

5. Feeding soil life by managing organic matter.

BIOLOGICAL BALANCE

OUT OF BALANCE	IN BALANCE
Anaerobic — decay without air	Aerobic — decay with air

SYMPTOMS:	RESULTS:
1. Little decay of organic matter.	1. Rapid decay of organic matter.
2. Sour smell; anaerobic decay which produces alcohol and formaldehyde.	2. Loose soil with earthy smell.
3. Insect and disease problems.	3. Earthworms present.
	4. Better crop growth.
	5. Healthier plants.

CAUSES:	REASONS:
1. Soil fertility out of balance.	1. Use of life-promoting fertilizers to get balance.
2. Excessive nitrogen use.	2. Abundant soil organisms
3. Working soil with wrong tools at wrong time.	3. Proper tillage — manage air and water.
4. Poor organic matter management (till too deep, too much).	4. Feeding soil life by managing organic matter.
5. Too many toxic materials; soil life is harmed.	5. High humus content.

CHEMICAL BALANCE

OUT OF BALANCE	IN BALANCE
Calcium less than 65% of CEC.	Calcium 70-85% of CEC.
Magnesium over 20% of CEC.	Magnesium 12-18% of CEC.
Potassium less than 3% of CEC or more than 5%.	Potassium 3-5% of CEC.
Phosphorus less than 20 ppm (P1).	Phosphorus 50 ppm (P1).
Sulfur less than 20 ppm.	Sulfur over 25 ppm.
N:S ratio over 15:1.	N:S ratio 10:1.
pH less than 6.0, over 7.0.	pH 6.5-6.8.
Low OM (organic matter).	Medium to high OM.
Low trace elements.	Adequate trace elements.

SYMPTOMS:	RESULTS:
1. Hollow stems (alfalfa), difficult to establish, short-lived stands.	1. Solid stems (alfalfa), easy-to-establish, long-lived stands.
2. Poor dry-down of crops.	2. Good dry-down & keeping quality.
3. Low sugar content in plant.	3. High sugar content.
4. Mineral imbalance in feed.	4. Good mineral balance.
5. Herd health problems.	5. Healthy animals.
6. Crops stressed by weather.	6. High yield; low weather stress.
7. Weed problems.	7. Few weeds.

Releasing Soil Nutrient Reserves

Plants require more of some nutrients than others (see table in Chapter 5, "Typical Plant Composition"). Typical plants take up much more nitrogen, potassium, calcium, sulfur, magnesium and phosphorus than they do manganese, iron, zinc, chlorine, boron, copper or molybdenum. That is why the terms "major," "secondary" and "trace elements" (or "micronutrients") are used.

What matters is how much gets into the plant, not how much is in the soil. As we have seen in Chapter 4, the soil has far larger reserves of most nutrients than plants need, but most of the total is unavailable at one time (see table in Chapter 4, "Composition of Typical Soils"). Most soils have several thousand pounds per acre of nitrogen, phosphorus, calcium, magnesium and sulfur (in the plow layer), plus about 30,000 pounds per acre of potassium. There can be about 50 to 200 pounds per acre of most trace elements.

Yearly crop needs are much less than the total nutrient content of the soil, typically about 0.5 to 2 percent. The amount of a nutrient that is readily available or exchangeable to plants each year is also only a small fraction of the total nutrient content, typically from less than 1 percent to over 30 percent (about 5 percent would be a rough average). For some nutrients the yearly crop needs exceed the available amount (according to textbook figures). The conventional crop fertilization philosophy assumes that soils may not supply all the crop's needs, and tries to add back the nutrients the crop uses, but only replaces P-K. What about calcium, sulfur, zinc, etc.? This idea treats soil as a solid medium to support the roots, while a nutrient solution of N-P-K floods the roots — basically hydroponic agriculture.

But the conventional philosophy doesn't take into account that some of the unavailable soil nutrient reserves can be made available each year through several natural processes: (1) weathering of soil minerals, (2) decay of organic matter by microorganisms, and (3) breakdown of soil minerals and release of tightly held nutrients by acids and chelating agents from plant roots, microorganisms and humus. The processing of soil by earthworms' digestive systems also helps, but most of that action occurs from microorganisms.

Note the significant role of microorganisms and organic matter. This shows the importance of regularly recycling organic matter back to the soil, of promoting high populations of soil organisms and of maintaining high levels of humus.

Through these natural processes and wise farming methods, a balanced "living" soil can, with added food (organic matter) for the organisms, supply all or nearly all of the yearly needs of most crops. This assumes that you have a soil type that has an adequate reserve supply of all essential nutrients (and is not deficient in a trace element, for example) and has an adequate cation exchange capacity to be able to hold the nutrients that do become available from the reserves (and thus keep them from leaching). That is why low-CEC soils such as sand cannot usually grow a good crop without supplementary fertilizers. But as many biological farmers have found, after several years their annual needs for outside fertilizers decline markedly, while available nutrients on their soil test reports keep climbing. The "extra" nutrients did not appear by magic, and not all came from recycled organic matter. Some came from the soil's huge reserve supply.

The relative amounts of reserve nutrients, exchangeable nutrients (held on soil colloids), readily available nutrients (in soil solution), and crop nutrient needs can be seen in the following diagram, where the different yearly nutrient amounts are represented as storage tanks with pipes running between them. The size of the pipes shows the rate of flow from one nutrient source to another.

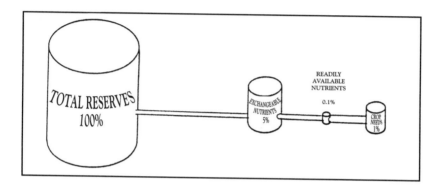

(The above proportions hold for phosphorus, potassium and magnesium.)

Unfortunately, it is difficult to know for sure in many cases exactly how large the soil's total — or available — nutrient supply really is. A soil test generally measures the exchangeable and readily available amounts (except for the P2 test, which is a measure of phosphorus reserves). Thus a typical soil test tries to estimate the amount available to a crop in one growing season (but there are different kinds of tests; see Chapter 8).

However, a soil test only tells you part of the story. Unless you rely entirely on outside soluble fertilizers, you should also know how effective the soil's natural release system is (shown by healthy crop roots and high levels of humus and soil organisms). You need to give nutrient credits for previous legume crops, green manures and other added organic matter. There are "textbook" estimates for these, but there are no good estimates of additional nutrients from soil minerals by root and microbial activity; it would depend on soil structure and aeration, weather, etc.

Another type of test often used to monitor nutrient effectiveness is plant tissue testing, where the actual amounts of nutrients taken up by the plant are measured. This can be a useful check. If soil testing shows sufficient levels of all available nutrients and plant usage is low or deficient, then you have to consider whether root health and absorption are adequate, or perhaps whether there is an imbalance of nutrients or a soil pH problem that reduces absorption of one or more elements (see below).

You can find descriptions or pictures of nutrient deficiency symptoms as they affect plant leaves and growth characteristics, such as yellow or reddish leaves or stunted top growth. These symptoms generally only develop after there is a severe deficiency, so it may be difficult to correct the deficiency that season. Foliar feeding may help the crop, but won't correct the soil. It is better to monitor the soil with regular soil testing and to feed the crop a balanced diet, rather than waiting until serious deficiencies occur. Another problem is that there can be more than one element that is deficient, making symptoms difficult to interpret. Also, because of nutrient interactions, the real problem may be that the level of an element is too high (see below).

Limiting Elements

Since all of the essential nutrients are needed by plants (not just N, P and K) and they are needed in a certain proportion or balance (different species of plants may need different proportions), it follows that *the productivity of a soil can never be greater than the plant nutrient in least supply.* The element in least supply is the crop's first limiting factor, nutritionally speaking (there could also be limiting factors from poor soil structure or wrong forms of microorganisms). The element in second lowest supply is the next limiting factor, and so on. Limiting factors hold you back. They reduce your yields below the crop's genetic potential. They reduce crop quality, which can lead to sick animals or people. This concept can be visualized by a barrel with staves of different lengths. You can't fill the barrel (yield, quality) any higher than the shortest stave allows.

Nutrient Interactions

Soil nutrients are related to one another. Soil chemistry is complex. An excess of one element can cause a deficiency of another. Some nutrient interactions and effects on crops include:

- Excessive nitrogen makes poor quality feed (nitrates).
- Excessive nitrogen (ammonia) can "burn up" humus.
- Excessive nitrogen can cause a potassium deficiency.
- Excessive magnesium can cause a potassium, phosphorus and nitrogen deficiency.
- Excessive potassium, sodium and magnesium can cause a calcium deficiency.
- Excessive calcium can cause phosphorus and trace element deficiencies.
- Excessive manganese can cause effects similar to iron deficiency.
- Excessive boron can cause signs of potassium and magnesium deficiencies.
- Excessive sodium and/or chlorine can cause signs of potassium deficiency.

The other factor that influences nutrient availability and interactions is the soil's pH, or acidity and alkalinity. Many elements have reduced or increased availability at too low or too high pH; that is partly why the "ideal" pH range is usually given as slightly acid, approximately from 6.0 to 6.8. Too much of certain elements can be toxic to plants, such as aluminum, manganese, boron, copper, chlorine and some "heavy metals" (cadmium, mercury, etc.). Note how the pH affects the availability of the following nutrients, as shown by the widths of the bands:

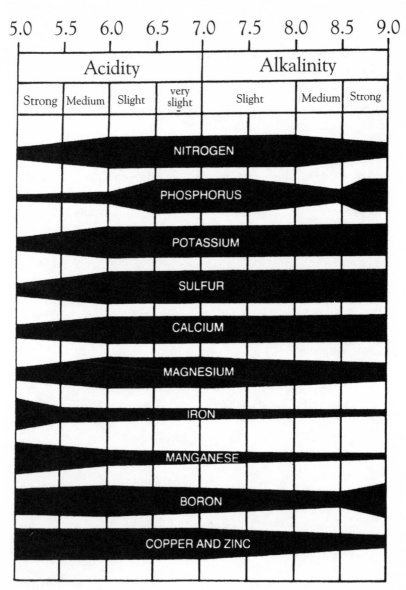

Availability of Elements to Plants at Different pH Levels for Typical Soils

Nutrient Flow Cycle

Farming is not a "numbers game," meaning that if you put so much on, you grow so much crop. Your soils have a great potential to convert nutrients already present into usable (available) forms by the aid of soil organisms. Without soil life to convert these minerals, you have to put on more fertilizer to get a crop. As soil life increases, the fertilizer needed to grow the same crop goes down. This is one way biological agriculture is so profitable.

We can diagram the main changes that occur and the flow of nutrients in a farm ecosystem as follows. Note the importance of soil organisms. Of course, the actual number of steps and changes in the soil are much more complex than shown.

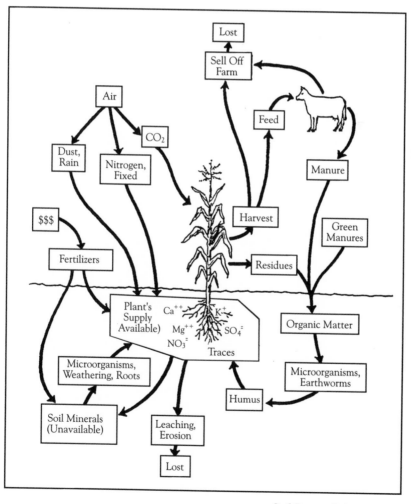

7

EVALUATION: HOW DO MY SOILS RATE?

You have been given a lot of base information. The *whats* and *whys*. Now I will start including the *how-tos*. This is not formula farming. I can give specific guidelines to follow and I will use case examples of workable farm programs. If you are still confused or are not convinced, contact one of our consultants and let them help you put together a working program.

To start any successful evaluation, you have to look at the big picture and get all the cards on the table before you begin. If you know where you are, know where you are going, know your limitations, and move forward, taking action will probably get you there. The biggest problem I see with farmers and biological farming is that farmers think they can purchase a product and things will work great without changing their basic farming practices. When they see no results or don't know how to evaluate their progress, they say, "Oh, this doesn't work." In reality, it does work. There are many, many successful biological farmers and many more farmers who are making positive changes toward that goal.

Step one is your mind — your thinking. You have to check it out and convince yourself there is something to this thing called biological farming. You can do this by reading, studying, going to meetings, visiting some successful farms, and following your common sense. But don't copy what these farmers purchase for inputs. Look at their practices; their farms are not yours. They have different soils, different farming histories, and different tools and products were used. The question is, how do *you* get your farm to work? What do *you* need to change? What are *your* limiting factors? If someone talks about a miracle product, or doing just this one thing that will make everything great — it probably won't work. What are the chances that one product will take care of all of your limiting factors?

There are certain practices that successful biological farmers follow and certain things they avoid. There is no one single exact way — all soils are different. Farmers perceive things differently and have different resources. They are in different spots, whether by accident or from a successfully planned practice. In an earlier chapter, I talked about the "rules" of biological farming — study them. How many are you following? You might be able to deviate from them, but to totally violate the rules and still be successful biologically is something I have never observed.

To start this soil evaluation, I will divide the soil into its parts: chemical (nutrients), physical (structure or tilth) and biological (organic matter and soil life). Now let's look at the history of what has been done to this soil and how the crops have been doing.

I. History

Chemical —

- Is there a complete soil test (organic matter, CEC, major elements, trace elements, base saturation)?
- Liming practices — has it been done? Type of lime used?
    - Was liming based on pH or on nutrient needed?
- Fertilizer practices — only N-P-K used? The cheapest source?
    - Other fertilizer materials used — sources?
    - Amount of nitrogen used — source?
    - Manure or compost used — how — amount?
- Weed and insect control — are insecticides used?
- Herbicides used — type — amount?
    - Weed types and pressure?
    - Insect pressure and species?

Physical —

- Tillage practices — type of tillage — what tools — when — amount?
    - Has subsoiling been done?
    - Are row-crops cultivated?
    - Are there row units or tillage practices done at planting?
    - How many trips over the field?

- Is compaction watched or a concern — would you drive on the field when too wet?
- Crop management practices — Are residues incorporated into soil — chopped in small pieces?
  - Are only row-crops grown?
  - Are crops rotated?
  - Any livestock manure — how is it used?
  - Any green manure crops?

Biological —

- Soil life — are earthworms present? Have there been earthworms?
  - Does crop residue decay rapidly and completely?
  - Have there been any soil biological tests?
- Organic matter — are residues present deep in the soil? Are they moldy?
  - Has the soil always had a nice earthy smell (like a root cellar)?
- Other signs — are there insect and disease problems with the crops? What and when?
  - Have weed types changed over the years — what major weeds are now present?

Crops —

- Crops grown — yield — patterns over time?
  - Fertilizer use for yield?
  - Crop problems?

All this information gives you clues to where you are. If we now do a present-time evaluation — testing, digging, and observing crop and soil performance — it will be easier starting or advancing in biological farming.

II. Current Soil Characteristics

Chemical —

- Get a complete soil audit — Look at nutrient levels and ratios among nutrients
  - Note excesses as well as deficiencies.
  - Evaluate soil capacity — Obtain your soil type maps — Use soil CEC (cation exchange capacity) as a base — What level of organic matter or humus?
- Soil type — What are the soil's limits? Is drainage a problem? Is moisture and water-holding capacity satisfactory?

Physical —

- Soil structure or tilth — is it loose and crumbly? Do chunks crumble with ease? Does it probe easily? Are there hardpans or surface crust? Is it sticky when wet?
- Other signs — are plant roots healthy, large and growing without restrictions?

Biological —

- Count earthworms (spring and fall are the most ideal times).
- Smell the soil.
- Observe the residue decay cycle — Are residues decaying properly and quickly?
- Check insect populations, both good and bad.
- Observe plant diseases — types and amount.

III. Testing

Testing is what testing is. Your objective is not to have ideal numbers from the test, but to use the test to give you clues as to how the soil is performing. With practical observation and common sense, you can make changes in your farming program leading to sustainable, profitable biological farming.

The list of test types is long, from soils to plant tissues, refractometers, chlorophyll meters, ergs test, penetrometer, soil microbiology, paramagnetic test, humus test, and probably many more. Don't get hung up on the results of any one test. There are seasonal variations, weather variations, and crop rotation variations. Testing is also a way of monitoring what is happening — a benchmark that allows progress to be noted. If you have livestock, the feed test plus the health and performance of livestock give clues. Ideal feed does not solve all livestock farmers' problems. Feed is just one thing to get right, and that's what this evaluation is all about.

Soil nutrient testing is the most often used method of testing and has the most information available about its usefulness. Soil tests use an extraction method to measure nutrients that are assumed a crop can get during a growing season. Different testing labs use different methods. They do not measure total minerals in a soil. That is why sending samples to different labs gives different results. You have to know how the lab interprets the results to be able to get any meaning from the results.

The secret to biological farming is using crops, soil life and fertilizers to make more minerals — currently tied up in the soil — available to future crops. There are many more minerals in a soil than the test measures and that is why good biological farmers see soil improvement. Soil tests indicate that mineral levels increase in the soil without adding more fertilizers so farmers, therefore, can reduce inputs.

What soil tests measure:
- Nutrients assumed to be useable by a crop
- Organic matter
- pH
- Ratios and balance of nutrients

Soil tests do not measure:
- Nutrient availability
- Biological activity
- Tilth or soil structure
- Minerals that are in the residue decay cycle
- The farmer's management ability of the soils

Knowing what a soil test measures and what it doesn't is essential for putting together a working farm program. You have everything in perspective; you have an understanding of where you are and what the limitations are — now what?

The next five chapters give background information on taking soil tests, understanding the numbers, and lime and fertilizer materials commonly available. Read and study these materials, and then we will be ready to look at case studies of farms and put biological farming practices in place.

8

SOIL TESTING: WHAT & HOW

Soil is such a complex and dynamic substance that it is very hard to measure or describe — sort of like the old story of several blind men trying to describe an elephant. Soil scientists and agronomists have devised many different tests that measure some particular aspect of the soil — just like feeling the elephant's ear, or tail. One test or even a few tests don't come close to really describing the soil at any one time. Add to that the fact that soil characteristics are constantly changing and you can see the magnitude of the problem.

There are many different aspects of the soil that you can measure, but just measuring things to get a bunch of impressive looking numbers has no purpose unless testing helps you understand what your soil is like and what is happening in it. And there is more to growing good crops than having adequate levels of nutrients in the soil (see Chapter 6). As a standard textbook says:

> Soil analysis, however, is no panacea. It will not supply answers to unsatisfactory plant growth when the cause is dry weather, compacted soils, critically low or high temperatures, inadequate soil drainage and low oxygen in the root zone, improper placement of fertilizer, salt accumulation, plant diseases, toxic elements, insect damage, competition from weeds or tree roots, or untimely operations. These probable limiting factors must be inventoried before soil tests can be interpreted for a fertilizer recommenda-

tion. (R.L. Donahue, R.W. Miller & J.C. Shickluna, An Introduction to
Soils and Plant Growth, *5th ed., 1983, p. 295.)*

That statement reiterates our belief that successful farming requires
paying attention to your entire farming system, not just trying out one
popular new practice or some hopeful "magic in a gallon jug." Soil testing
is simply a tool, used to help you or your consultant diagnose any problems, and monitor conditions and changes over the years. Most people
think of a nutrient test when you mention "soil test," but as we will see,
there are many other ways to test your soil.

For practical farming purposes, it is not necessary to measure every one
of your soil's properties to five decimal places every year. A standard
"complete" nutrient test (see next section) taken every two or three years,
or just before planting a new seeding of hay, will tell quite a bit that we
need to know in order to make a fertilizer recommendation. But still, a
nutrient test only measures the soil's chemical properties, and there are
also the physical and biological aspects of the soil that are very important.
Also, there are different ways of doing a nutrient test. In this chapter, I
want to introduce you to several of the soil tests more commonly used in
biological agriculture today.

Sampling

One problem in doing soil tests is getting a truly representative sample,
or taking enough samples to give an accurate reading. Soil is so variable
that two samples taken a few yards apart may give very different results.
The same goes for different depths as well for as samples from in the row
versus between the rows. Remember, a one-pound soil sample is supposed
to represent perhaps 10 or 20 acres. For tests to give a good picture of a
field, the sample must be typical of the soil you want to test. The test
results you get can only be as good as the sample you send in to the lab.

Generally the best information comes from taking at least 20 subsamples from various typical parts of the field (that is, if you want an average
picture of the field; but if you are trying to test a specific portion of a field
or a problem area, then obviously you would only take subsamples from
that part of the field). Fields with areas having different fertilization or
liming histories or those with greatly differing soil types can be split up and
tested separately.

Generally, you are interested in sampling from the plow layer, the upper
six or seven inches, but sometimes tests of the subsoil are done for special
reasons, such as to measure nitrates or toxic elements. Because available
nutrients or other soil properties can change a lot during the year, it is best
to always sample at the same time of year and at approximately the same

soil moisture content each time you do tests. That way you can compare "apples with apples." Samples can be taken in the winter, but nutrient availability will be lower then. Spring or fall are usually the best times to sample.

Obtain aerial photos of your farm and make a map drawing if this has not already been done. Number all of your fields.

For "average" testing, take subsamples from all typical parts of a field, either by walking around in a random pattern or back and forth in a systematic grid pattern. If some areas of the field are greatly different, either avoid taking samples there if they are small sections, knobs, etc., or if they are large enough areas that you can fertilize them separately, then sample them separately. If practical, mark on a field map the approximate locations of the subsamples. If possible, use a stainless steel soil probe or soil auger to prevent contamination from other metals. A shovel can be used, but do not let the soil that is saved touch iron or your fingers. The soil from the 20 or more subsamples can be collected in a clean plastic pail. Then mix it thoroughly with a plastic or wood tool and put about one pound of this composite sample into a soil sample bag, which testing labs supply (if no "official" bag is available, a strong paper or plastic bag can be used). Then be sure to accurately and completely label the bag with the information the lab wants: usually your name, address, date and the field number or letter (maybe also the particular tests to be run and the crop to be grown). If you ridge-till or no-till or graze, note the depth of sampling and position on a ridge. Such soils also need to have shallower samples taken, since nutrients tend to concentrate near the top. If there is a large difference, some type of aeration may be necessary to move nutrients downward. If nutrients are near the top and the surface dries out, how can the crop perform?

Ways to Test Soil

Briefly, what follows are some ways to get at least a partial idea of what all three aspects of your soil are like: chemical (nutrients), physical (structure) and biological (life). Some tests give an indirect indication rather than a direct measure.

Nutrient Test, Mild Acid Extraction

This is the "standard" nutrient test that everyone thinks of as a "soil test" (see next section). Actually, there are several methods of testing for the various nutrient elements, and certain methods may give better results for certain types of soil. In general, a soil sample is dried, ground up to provide uniform-sized particles and soaked in a weak acid that dis-

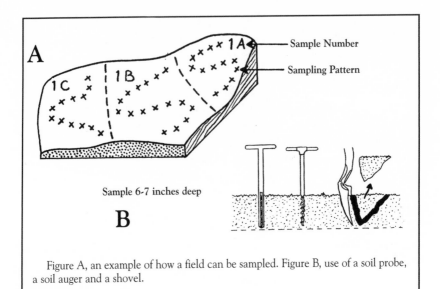

Figure A, an example of how a field can be sampled. Figure B, use of a soil probe, a soil auger and a shovel.

solves (extracts) a portion of the elements held by the soil particles, supposedly to simulate what nutrients the acids from plant roots will release over a growing season. Sometimes a salt solution is used as the extractant rather than an acid. You can find out the details of testing by asking the laboratory that performs this test.

The results are usually reported in parts per million (ppm) or sometimes pounds per acre (parts per million multiplied by two equals pounds per acre, since there are about two million pounds of soil in the plow layer). The report may also give indications of whether the nutrient supply is considered adequate for the crop to be grown typically by using words such as high, medium or low. Additionally, the soil's pH (acidity or alkalinity), CEC (cation exchange capacity) and organic matter content may be reported.

Nutrient Test, Water-Soluble

This is a test for nutrient elements in which the soil sample is only soaked in water, not mild acid. It is supposed to tell you how much nutrient is immediately available to plants — like an instant snapshot — rather than what plants may absorb all season. Therefore, the results in ppm or pounds/acre will be lower than in an acid extraction test. You will need to become used to reading (interpreting) a water-soluble test to get much good from it. Farmers need to run several water-soluble tests during the growing season. Some even do it on a weekly basis to get a better picture of what is happening.

Ergs Test

This is a specialized test that some agriculturists say measures the soil's "energy," or its ability to produce good crops. Actually, it is a measure of conductivity, and a standard conductivity meter is used (you can buy your own meter for as low as $40.00). The meter reads in millisiemens/meter (previously called micromhos/centimeter; also called "ergs"). Conductivity usually measures the concentration of salts in the soil. A mixture of soil and pure water is used for the test. Salt ions conduct electricity from the meter and cause the meter to give a reading. Most nutrients that are available to plants are in the form of salt ions such as sulfate, phosphate, nitrate, potassium, and so on. Usually, the higher the "ergs" reading the higher the available nutrients, and the higher the soil's "energy." Desired readings run from 100 to 800 ergs, depending on the crop and the time of year. Much above 800 may indicate a serious salt or salinity problem. High salt levels can injure or kill roots and seedlings, as well as soil life. Salts will rise and collect in the uppermost soil layer during a drought or after irrigation in dry weather. High soil salts are often a problem in the western United States, or anywhere high levels of salt fertilizers have been used. This test is usually run in conjunction with a water-soluble test.

Penetrometer Test

This is one type of measurement of the physical condition of the soil, or the soil structure. A loose, well-aerated soil is desirable. A penetrometer is an instrument with a pressure gauge attached to a metal rod that is pushed into the soil. The gauge reads the pressure (in pounds/square inch or kilograms per square centimeter) while you penetrate the soil. The reading will vary as you go deeper, allowing you to easily locate a hardpan which may restrict water and root penetration. To get accurate readings, have an assistant help you record gauge readings at three-inch intervals of depth; then take at least five to 10 readings at different places and average them. Readings should be taken when the soil is moist, for instance two days after a soaking rain. Below 150 pounds/square inch is excellent, while over 250 is poor. A penetrometer can be purchased for as low as $175.00, or you can just use a ½ inch steel rod to get a rough idea of your soil's looseness. Digging in the soil with a shovel can also give you an indication of soil structure and any compact layers.

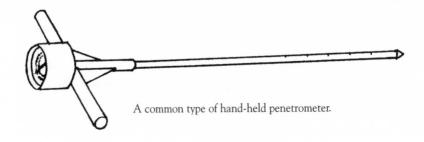

A common type of hand-held penetrometer.

Refractometer

Refractometer tests are only an indirect measure of the soil's condition and fertility. Actually, the refractometer measures the sugar content (technically, the soluble solids) of the juice of plants. Typically, a drop or two of sap from some leaves (or leaves plus stems, or stems) are squeezed onto the refractometer and the sugar content is read from a scale in units called Brix (equals percent sugar).

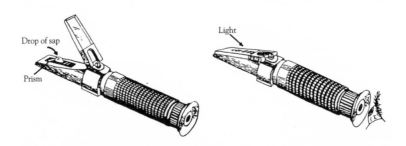

Drop of sap

Prism

Light

Higher sugar generally indicates a healthier plant growing on good, fertile soil. But there are many factors that can cause low sugar, including young plants, cool temperatures and cloudy weather (sugar is produced by photosynthesis in the leaves). You have to be careful to take measurements from the same part of the plants (leaves, stems or roots) each time, and at the same time of day in the same weather to get a good comparison since all of these affect photosynthesis or sugar content. Normally, readings from leaves will be highest at mid-day and lowest in early morning. High readings in morning could indicate the plant is "sick" and not transporting sugar from leaves to other parts the way it should. Brix readings below six are rated poor for most crops, and above 12 is excellent. Some people believe that when sugar readings are above 12 Brix, the crop will not be bothered by insect pests; there are indications that this holds true most of the time. A refractometer costs from $175.00 to over $200.00.

Earthworm Counts

Earthworms are good indicators of the health of the soil. They proliferate in soil that is high in calcium and has good structure, plenty of organic matter, and beneficial microorganisms. Earthworms are large enough to see compared to microorganisms which are microscopic. During the growing season (especially spring and fall), an earthworm population of 25 worms per cubic foot of soil indicates excellent soil condition, while below five per cubic foot is poor. Worm populations can be counted when the soil is moist (they are inactive in drought) by either counting worms in the soil or by counting their castings on the surface. You can dig out a cubic foot of soil and sift through it for worms, or you can clear off the plant residues and old castings from a square foot and then count the castings that appear on the surface a day or two later. To get accurate counts, repeat at least three times and take an average. You should realize that not all soils will have many earthworms, yet they may be good, fertile soils. Earthworms are simply rare or absent in some geographic areas, and sandy soils seldom have many.

Microorganism Tests

Because of their small size, the soil's microorganisms are difficult to count individually. Some labs can do a plate count, in which the colonies of bacteria and/or fungi that grow on a petri dish are counted, and the number of microorganisms per gram of soil is estimated. The desired situation is to have a large variety of species, with about a 1:1 ratio of fungi to bacteria. This test is fairly expensive. Dr. Elaine Ingham of Soil Foodweb Inc. in Oregon has been doing a lot of research and testing in this area. This test, as with a water-soluble nutrient test, varies greatly throughout a season, so conditions at testing need to be noted.

An indirect indication of soil microbial activity can be obtained by measuring the by-products they give off. One such measure is soil respiration, in which the carbon dioxide that accumulates from an incubated soil sample is measured. This is usually an expensive laboratory test, but it can be adapted to field use. Soil respiration varies with temperature and moisture. Low readings would be below 20 pounds of carbon dioxide per acre per day, and above 60 would be good. Very high readings may not be good, since it could indicate that microbes are "burning up" organic matter, as could happen when too much raw manure is applied.

Some labs offer a test of microbial enzyme production. One such test is sometimes called the formazan test. It measures the amount of a type of enzyme (formazan, a dehydrogenase enzyme) produced by certain soil bacteria, so it is only a partial measure of all types of microorganisms. The formazan test is reported by the lab in units called micrograms formazan/10 grams soil/day. Numbers below 100 are poor, and above 1,000 are excellent.

Various Nutrient Tests

Now let's look at the different types of soil nutrient tests that are available in the testing and consulting industry. Various fertilizer companies, dealers and testing labs may provide very different kinds of soil nutrient tests. They may or may not tell all we need to know to balance the soil.

P (Phosphorus), K (Potassium) and pH Test

P, K and pH tests are inexpensive or free from many fertilizer companies. But they only test for two elements, phosphorus and potassium (that's what the companies sell). What about the other 10 or more soil elements crops need? Many times, the test results come back high to very high, yet the salesman still recommends more P and K. Research and common sense both indicate this is incomplete testing and a wasteful way to fertilize.

Major Test

This tests for OM (organic matter), P (phosphorus), K (potassium), Ca (calcium), Mg (magnesium), S (sulfur), pH and CEC (cation exchange capacity). It is a good starting test — a minimum of what needs to be tested. It tells you something, but not everything you need to know. You can get a start at soil balancing, but you don't know enough to spot some possible problems.

Complete Test

This includes everything in the "major test," plus the trace elements: Zn (zinc), Mn (manganese), Fe (iron), Cu (copper) and B (boron). For special situations, or for soils in particular areas, a few additional tests may be run such as sodium, soluble salts, chloride, nitrate, hydrogen or others. This is a much better test and only costs a few dollars extra. It's well worthwhile and gives the information needed to begin balancing the soil.

Soil Testing Variables

There are various factors that can affect soil test results and the numbers on your report sheet. How accurate and reliable are soil tests? Are they wasted money?

No, soil tests are not useless, but you have to realize what they measure and what their limitations are. Soil tests measure the amount of extractable nutrients (or other quantity such as pH or organic matter) in that particular one-pound sample of soil. That's all. But how accurately do the results reflect what is actually in the field from which you took the sample? Do the numbers mean that the field will have those levels of nutrients there all year? There are several factors that influence the numbers on a soil test report.

Sampling

As we mentioned earlier, the test results are only as good as the sample. Soil scientists say a representative sample is the most critical step in soil testing. Yet how often do we just walk a few yards into a field, scoop up a shovel of soil and send it in?

Scientific studies of field variability often find large differences from one part of a supposedly uniform field to another. When mapped out, it may look like a patchwork quilt. Just consider the differences in fertilization that may occur from in-row or band applications. Some researchers and companies are developing highly accurate testing and fertilizer application equipment that custom treats small plots at a time (about 200 to 450 feet square). The initial soil testing costs are very high, and there can still be field variability within the one- to 4.5-acre plots.

Obviously, if you intend to treat a field uniformly, you need to get an average estimate of soil nutrients. So, ignoring small unusual areas, such as low, wet spots, droughty sand knolls, etc., you should take subsamples from all parts of a field, then mix them thoroughly to get a representative composite sample. Typically, at least 20 cores or subsamples are needed to get an accurate test. For most fields with very different soil types, 30 cores would be better; in very uniform fields, 15 cores will suffice.

You need to either sample between the rows or in the rows, depending on what you want to know or how you plan to fertilize (general soil correction versus row-support fertilization).

Also consider the effect of depth and the possibility of stratification, or layering, of soil. Tillage tends to mix up soil and distribute nutrients somewhat uniformly, while no-till agriculture, with surface-application of lime and fertilizers, leads to much different fertility levels and pH in the top few inches compared to lower down.

Seasonal Variability

If you ran soil tests throughout the year, you would find that the pH and available nutrient levels can change considerably. Typically, the pH declines throughout the growing season (becomes more acid), as the crop takes up exchangeable nutrients and replaces them with hydrogen (see Chapter 5). Some available nutrients increase and peak in the warmest part of the season, but often decline later as crops use up the available supply or as the leachable ones leach away. Soil test levels also depend on soil moisture; a drought can lower the pH, phosphorus and potassium readings. High or low pHs affect the availability of a number of elements, such as phosphorus and trace elements (see Chapter 6). Microbial activity and organic matter breakdown affect the amount of nitrogen, phosphorus, sulfur and some trace elements. This seasonal variability is why we recommend that you take soil samples at about the same time each year.

Soil Chemistry

We just mentioned that soil pH affects the availability of some elements. Other ways soil chemistry can affect soil tests is through the interaction of certain elements, such as calcium, magnesium and potassium. The total amount of all three that soil colloids can hold is a certain level which depends on the soil's cation exchange capacity (see Chapter 5). If one of these elements is very high, such as magnesium or potassium, the levels of the other two will automatically be lower. High levels of aluminum, iron and sodium can cause large (and detrimental) changes in nutrient availability or can even be toxic to plants.

Differences in Testing

Exactly how the testing lab does the test can result in different values. That is partly why some labs report results in terms of "high," "medium" and "low" rather than numbers — to prevent confusion. Most labs dry the sample in an oven and grind it to get uniform particle size, but these operations can change the pH or nutrient availability from what they were in your field. Still, by doing this, the lab can give reliable results for dry, ground-up soil. A few labs will run tests on naturally moist, "as-is" soil. Since the results will be different from the usual procedure, the person who interprets the test has to make adjustments.

The nutrient extraction method can vary in different labs or for different elements. They can use water alone, a salt solution, or a mild acid solution to extract a portion of the sample's nutrients. The sample can be shaken for varying lengths of time (the longer it is shaken, the more nutrients are extracted).

Then there are many chemical test procedures that can be used. Some labs prefer one over another because they believe the results are more reliable for the soils in their area. For example, most labs prefer a water-extracted pH reading over a salt pH or a buffer pH test, or there can be a Bray P1 test for phosphorus, or a Bray P2, or a sodium bicarbonate extraction, or a Mehlich acid extraction. All these testing procedures measure phosphorus, but they all give *different* numbers from the same sample. Soil scientists are trying to set more uniform standard soil testing procedures for the entire United States.

All the different possible tests and results are pretty confusing to the layman. How are they interpreted? When a soil test method is first developed, scientists run field calibration tests on soils of different nutrient levels. They compare the soil test results and the growth of crops (usually dry matter yield). That is how they work out the "low," "medium" and "high" ranges.

Finally, another problem is that it is simply very difficult to test for some soil nutrients, or at least a really reliable test has not yet been developed. In other words, the numbers from a test don't always closely reflect the nutrients a plant's roots can take up. The tests with the most variability for this reason are for iron, molybdenum, nitrogen, manganese, copper, sulfur, calcium, magnesium and boron.

So, to come back to our original questions, are soil tests really accurate or worth the money? We have seen that they aren't totally accurate, if you mean to three decimal places. You shouldn't be dazzled by seemingly accurate numbers or be concerned by year-to-year differences of a few percent of the total amount. Sampling differences, field variability and seasonal differences can introduce a certain amount of error into soil testing.

But don't throw the baby out with the bath water. Imperfect as they are, soil tests are the best estimate we have of available soil nutrients. They give us at least an indication of what is going on. They are much better than "flying by the seat of your pants." How can you even begin to correct an unbalanced soil if you have no idea what is out of balance?

9

SOIL TESTING: SOIL REPORTS

Different soil testing labs send back the results in somewhat different ways. Some labs give only the numbers — parts per million (ppm) or pounds per acre of nutrients, plus other things like percent OM (percent organic matter), CEC (cation exchange capacity) and percent base saturation (percent of nutrient cations on soil colloids). Other labs only give relative sufficiency levels — very high, high, medium, low and very low, perhaps also with graphs of the actual amounts. Some labs may give both actual numbers and sufficiency levels. You need to study reports from the lab you use to become familiar with their reporting system. The exact order in which the various items are presented on the page will also be different. Some labs or consultants may also add "desired" or "ideal" levels of each nutrient so you can see whether you are too low or too high, and by how much your soils are too low or too high.

The purpose of soil testing is to give you clues as to what is happening in the soil. It is a long way from being an exact science. You are looking for trends and ratios, not exact numbers. If the test indicates your soil is low in a nutrient, you need to add it. If it is high, don't put any more on. It is really that simple; you are balancing the soil nutrients. How much you add and how fast you correct the problem depends on many factors such as money, crop to be grown, etc.

In working toward balancing soil nutrients, if pH and/or calcium are low, and phosphorus and potassium are low, you need to start there. If you have livestock, a major reason to test the soil is to know where to spread the manure. You want all fields to be in the medium to high range. This will give you better overall crops and a uniform feed supply for better livestock health and performance.

Two things are at work here. One is soil balance; the other is crop fertilizers. The crop fertilizer can be adjusted to make up for soil deficiencies while you are in the process of balancing. The fertilizer should always be balanced for the growing crop, with major, secondary and trace elements.

How to Read a Soil Report

Let's go through the typical items on a hypothetical soil report.

Parts Per Million (ppm) or Pounds Per Acre

An acre of topsoil six to seven inches deep weighs about 2,000,000 pounds, so one ppm = two pounds/acre. Multiply parts per million by two to get pounds per acre, or divide pounds per acre by two to get parts per million.

Organic Matter (OM)

A measure of plant and animal residues in the soil, including both raw and decomposed organic matter. Some labs offer a separate test for percent humus, which measures only the decomposed organic matter. The humus provides many valuable functions in the soil. It is related to the soil's color. Darker soils are usually higher in organic matter. I like to see at least 2 and up to 5 percent OM. Not using excessive nitrogen applications and adding plant and animal matter (such as corn stalks, manure or cover crops) will raise soil organic matter levels.

Some labs will also give Estimated Nitrogen Release (ENR), which is calculated from the percent OM. It is supposed to be the amount of nitrogen that should be released from the soil by microbial action through the growing season. But that is all hypothetical since actual nitrogen release depends on the type of organic matter (humus versus raw organic matter), soil conditions (aerated versus waterlogged), and weather (temperature, moisture). If a heavy rain leaches away most of your nitrogen, you will probably need to side dress more. There is no "ideal" level for ENR, and it is not a very useful number for the "real world." You can grow your own nitrogen (legumes, organic matter, microorganisms), but you can't grow mineral nutrients. Therefore, I don't emphasize measuring soil nitrogen, but I try to provide it as needed to the crop, only using purchased synthetic nitrogen as a last resort.

Cation Exchange Capacity (CEC)

This is a measure of the soil's ability to hold nutrients. The higher the CEC, the greater the soil's holding capacity. Colloidal soil particles (clay and humus) hold the nutrient ions. Sandy soils have a lower CEC and require more frequent nutrient additions than heavier soils. The CEC can be increased by adding organic matter. Organic matter holds both nutrients and water (see Chapter 5 for more details on CEC). There is no "desired" CEC — it is what it is. Generally, a CEC below 10 is low and over 25 or 30 is higher than average. Low-CEC soils will require supplemental fertilization to grow high- yielding crops.

Percent Base Saturation

This is the relative amount of cations (positively charged ions, or bases) held on soil colloids, expressed as a percent (they all total 100 percent; it is calculated from the CEC). It includes the three nutrient cations calcium, magnesium and potassium, plus hydrogen and sodium if they are present. The desired levels are 70 to 85 percent calcium, 12 to 18 percent magnesium, 3 to 5 percent potassium and less than 5 percent hydrogen. Ideally, sodium should be zero since it is not an essential plant nutrient, but many soils in the western United States will have high levels. I am speaking here primarily of temperate climate soils; many tropical soils, where high rainfall causes severe leaching, will have a different "best" balance.

You should realize that the percentages, or proportions, of nutrient cations could be "ideal," but the soil could still not provide them in sufficient quantity to the crop. This is because in a low CEC soil, there is not a large enough capacity to hold what the crop needs. This means the farmer will have to supplement low CEC soils with extra fertilizer. See the following table.

Desired Cation Levels

The following table gives desired soil test levels in parts per million (ppm) for potassium, magnesium and calcium at the average desired base saturation percentages of 3.2 percent for potassium, 15 percent for magnesium and 75 percent for calcium. These levels have been calculated for cation exchange capacities (CEC) from 4 to 20 in half steps. The formulas for calculation have been provided so calculation at other CECs or percentages can be done.

Desired Cation Levels

CEC	ppm Potassium 3.2% base saturation	ppm Magnesium 15% base saturation	ppm Calcium 75% base saturation
4.0	50*	72	600*
4.5	56*	81	675*
5.0	62*	90	750*
5.5	69*	99	825*
6.0	75*	108	900*
6.5	81*	117	975*
7.0	87*	126	1,050*
7.5	94*	135	1,125*
8.0	100*	144	1,200*
8.5	106*	153	1,275
9.0	112*	162	1,350
9.5	119*	171	1,425
10.0	125	180	1,500
10.5	131	189	1,575
11.0	137	198	1,650
11.5	144	207	1,725
12.0	150	216	1,800
12.5	156	225	1,875
13.0	162	234	1,950
13.5	168	243	2,025
14.0	175	252	2,100
14.5	181	261	2,175
15.0	187	270	2,250
15.5	193	279	2,325
16.0	200	288	2,400
16.5	206	297	2,475
17.0	212	306	2,550
17.5	218	315	2,625
18.0	225	324	2,700
18.5	231	333	2,775
19.0	237	342	2,850
19.5	243	351	2,925
20.0	250	360	3,000

Formulas: For Potassium: 780 x CEC x 3.2 ÷ 200 =
For Magnesium: 240 x CEC x 15 ÷ 200 =
For Calcium: 400 x CEC x 75 ÷ 200 =

* Low-CEC soils may need a higher level of potassium, calcium and magnesium than given in this table to produce high quality crops.

pH

This is a measure of the acidity or alkalinity of the soil, caused by hydrogen ions. A pH of seven is neutral. Anything below seven is acidic, and above seven is alkaline. Any pH that is a whole number above or below another has 10 times fewer or greater hydrogen ions (it is a logarithmic scale); for example, pH 5.4 has 10 times as many hydrogen ions as pH 6.4. Soil pHs between 6.2 and 6.8 are "ideal" for most crops. The hydrogen percent saturation is zero when the pH is seven. In an acid soil, hydrogen has replaced calcium, magnesium, potassium and sodium on soil colloids. The first three elements should be replenished in proper proportions (see above, "Percent Base Saturation").

Potassium (K)

A measure of readily available potassium (much more tied-up or unavailable potassium is present in soils). Desired levels: light soils = 125 ppm or 3.5 percent base saturation, medium soils = 150 ppm or 3.0 percent base saturation, and heavy soils = 175 ppm or 3.0 percent base saturation. High-magnesium soils above 20 percent base saturation require extra potassium.

Magnesium (Mg)

Soils in dolomitic bedrock areas may have a problem with too much magnesium, while light sandy soils may have too little. The desired range is 100 to 250 ppm, or 12 to 18 percent base saturation. A 2:1 ratio of magnesium to potassium (in parts per million or pounds per acre) is good.

Calcium (Ca)

Calcium is the nutrient needed at high levels for both plants and beneficial soil organisms. It improves soil structure and helps bring the balance of other elements into line. Calcium is a necessary plant nutrient, so even if pH is high, additional available calcium may be required. In high pH soils, sulfur or sulfate should be added with calcium. Desired levels are 1,500 to 2,000 ppm (or higher), or 70 to 85 percent calcium base saturation. There is some controversy over the balance of calcium and magnesium. I like to see a calcium to magnesium ratio of 5:1 to 7:1 in parts per million or pounds per acre (see Chapter 10 for more details).

Sodium (Na)

A measure of this element is normally only made on soils classified as sodic (high-sodium) or saline-sodic, such as are found in the western United States. High sodium in soils destroys soil structure by dispersing the small particles, and some crops are harmed by high sodium.

Phosphorus (P)

There are two tests usually done for phosphorus

P1 (weak Bray extraction). This is what is readily available to a plant now. Twenty-five ppm is a minimum level on soils with 3 percent organic matter, but 50 ppm would be ideal. At that level, except for corn starters, specialty crop requirements, and to "winterize" alfalfa, no additional phosphorus would be required.

P2 (strong Bray extraction). This measures the readily available phosphorus plus a part of the active reserve phosphorus, which should become available later in the season. Fifty ppm is a minimum level, while going to 100 ppm adds more crop quality and health. The ratio of P1:P2 should be 1:2. Wrong calcium levels and a pH too high or too low can change this ratio, indicating a soil condition for poor crop performance.

Sulfur (S)

Sulfur is a much-neglected nutrient. More sulfur than phosphorus is needed to grow some crops, and sulfur is essential for complete proteins. The sulfur-to-nitrogen ratio is very important since sulfur improves nitrogen availability. Twenty-five ppm is a good minimum soil test level, with a S:N ratio of 1:10 in the feeds.

Trace Elements (Micronutrients)

These are needed in small amounts. They can be effectively soil-applied or foliar fed. "Natural base" fertilizers contain trace elements, while most commercial N-P-K fertilizers have few if any. All crop fertilizers should have a balance of all the trace elements.

Desired levels:

Zinc (Zn) — 5 ppm
Manganese (Mn) — 20 ppm
Iron (Fe) — 20 ppm
Copper (Cu) — 2 ppm
Boron (B) — 2 ppm

Other trace elements, such as molybdenum and chlorine, are also needed by plants, but they are not normally deficient so laboratories usually do not test for them. Using a natural base fertilizer will provide some. Idaho phosphate is an excellent source of molybdenum.

Functions and Benefits of Elements

The benefits of some elements follow to show you how important they are. To get all of them to the plant they have to be in the soil in an exchangeable form, but many are in "competition" with each other, and the balance among them is critical. In fertilizing and balancing a soil, remember to watch the excesses and well as the deficiencies.

Phosphorus — photosynthesis, fast and vigorous growth, good and early root growth, energy release in cells, cell division and enlargement, a part of cells' DNA, increased nitrogen uptake, increased mineral and sugar content of plant, earlier maturity.

Potassium — stalk strength, lodging resistance, winter hardiness, disease resistance, protein production, carbohydrate production, sugar translocation, enzyme functions, cell division.

Calcium — improved soil structure, stimulates soil microbes and earthworms, mobilizes nutrients into plant, increased nitrogen utilization, higher protein content, root and leaf growth, cell division, stronger cell walls, enzyme functions, increased sugar content of plant, overall plant health, high quality grain or fruit.

Magnesium — is a part of chlorophyll, protein production, enzyme functions, energy release in cells, aids phosphorus uptake, oil formation, starch translocation.

Sulfur — loosens and aerates soil, reduces excess soil magnesium, lowers soil pH, makes soil nitrogen more available, more useable protein (high quality, complete), energy release in cells, part of vitamin B1 and biotin.

Zinc — contributes to test weight, increased corn ear size, promotes corn silking, hastens maturity, chlorophyll formation, enzyme functions, regulates plant growth.

Manganese — normal growth and photosynthesis, oil production, energy release in cells, enzyme functions.

Iron — chlorophyll production, energy release in cells, needed by nitrogen-fixing bacteria.

Boron — promotes flowering and pollen production, seed development, root and leaf growth, cell wall formation, protein production, sugar translocation, energy release in cells, improved crop quality, increased calcium uptake.

Soil Test Results and Recommendations

The numbers on a soil test report need to be interpreted in light of the particular farm's soil types, topography (slope), weather, crops and rotations, tillage practices, soil structure, organic matter management and biological activity. Then someone needs to decide how much of what kinds of fertilizers are needed. If the farmer doesn't have the time or expertise, then a consultant or fertilizer salesman usually makes recommendations. Following are a few principles and comments.

Soil testing is a guideline. It indicates (1) problems that fields may have; (2) what direction to go with applied nutrients, if any are needed at all; and (3) serves as a guide for types and amounts of fertilizers to apply. Soil testing is not an absolute measurement of actual soil fertility.

Soil is a "living" substance with minerals, organic matter, organisms, air and water. Soils are not mathematical and correcting them is not an exact science. Putting on 200 pounds per acre of a material does not change your soil test by 200 pounds. Remember, an acre of soil six to seven inches deep weighs two million pounds. The soil test estimates what levels of nutrients should be available during the growing season.

I would much rather have a soil with a fairly good balance of nutrients in the medium test range, with lots of biological activity and a good structure than a structureless, dead, high-testing soil. In a soil that tests excessively high in one or two nutrients (perhaps magnesium and potassium), crop performance is not what it should be. Crop quality is down (as measured by livestock health), and also the health of the soil and plants is poor. That is why insects are after the crop and certain weeds are out of control. It's nature's way of dealing with the problem.

Soil balance is not achieved with one application, one time. It has to be practical. Getting a good crop while working toward soil balance makes sense. If the major nutrients (phosphorus, potassium, calcium, magnesium, nitrogen and sulfur) are deficient or out of balance, that is the place to start.

Trace elements can be applied for that year's crop. It is a mistake to bulk spread a trace element to correct a soil deficiency until the major elements are in line. As major elements adjust toward balance, the whole soil will change. Many nutrient levels will improve with better soil conditions. More microorganisms and more organic matter mean more nutrients released from the two million pounds and available for crop use. For many soils, the place to start seems to be with calcium, and the next important spot to look at is phosphorus. Not only is calcium low in many soils, it is the key to making some important physical and biological changes in the soil.

How fast a soil changes depends on the level of fertilizer input, manure use, green manure crops worked into the soil, type of crop grown and certainly the biological activity. Balancing your soil while managing all your resources and growing a good crop makes sense.

Monitoring Soil Fertility Changes: Two Examples

We often hear that when a farm goes on a biological program, the soil will begin to change and become more in balance. Can we document this?

Following are soil tests from two southern Wisconsin farms that have used biological products and methods for several years. When you compare their 1991 soil tests with older ones, you can see a number of changes in nutrient elements and in other soil properties — all heading in the right direction. The accompanying notes point out other significant changes on these farms.

Notes and Observations, Two Southern Wisconsin Farms

1. Neither of these farms has applied corrective phosphorus or potassium fertilizer since they began the biological program, yet their soil test levels of phosphorus and potassium have increased markedly. Both have used dairy manure, however.

2. Farm 1 started using gypsum as a soil corrective in 1986. They also used blended biological fertilizers from Midwestern Bio-Ag and applied 28 percent liquid nitrogen. Two years later, because their soil was starting to change, they switched to a kiln dust product (BioCal) and continued with Bio-Ag fertilizers (200 pounds/acre corn starter and 200-250 pounds/acre alfalfa fertilizer). These are balanced fertilizers and contain trace elements. In the beginning years complete soil tests were not run so there is no record of the trace elements at the start.

3. Farm 2 started out with kiln dust as a soil corrective and used Bio-Ag corn starter (200 pounds/acre) and 5-8-12-13 S plus trace fertilizer (200-300 pounds/acre) on alfalfa. At first they added a colloidal soft rock phosphate to livestock manure, but have now switched to a rock phosphate from Idaho. They use no additional nitrogen and dropped all chemical use (herbicides) in 1989. The farm is now certified organic. They still use kiln dust on alfalfa, and new seedings and have changed to 125 pounds/acre corn starter and 150 pounds/acre of alfalfa fertilizer (fertilizer inputs have dropped and profitability has risen).

4. These comparisons are not scientific in any way; that is, they are not based on replicated tests. Still, if these reports came from my farm, I would not only feel good that the program I was using was maintaining soil fertility and not using up my nutrients, but I also would be excited at all the positive changes: increased organic matter, increased CEC, and increased levels of all nutrients except percent magnesium and the amount of iron. These two elements were in excess before, so we would want to see them decrease.

In addition, the crop production on these two farms is excellent. Insect and disease problems are non-existent, alfalfa stands are not dying out and livestock production and health keep on improving. What will the next five or ten years bring?

Soil Tests from Two Southern Wisconsin Farms Using Biological Products and Methods:

FARM 1	% OM	CEC	%K	%Mg	%Ca	pH	K	Mg	Ca	P1	P2	S	Zn	Mn	Fe	Cu	B*
1986-87 av.	2.9	12.3	2.7	37	59	7.2	129	548	1468	31	47	14	2.4	NA	NA	NA	1.3
1991 aver.	3.3	14.5	3.1	36	61	7.3	173	635	1752	45	62	18	5.6	21	36	1.5	1.3
Differences	0.4	2.2	0.4	-1	2	0.1	44	87	284	14	15	4	3.2	—	—	—	—

FARM 2	% OM	CEC	%K	%Mg	%Ca	pH	K	Mg	Ca	P1	P2	S	Zn	Mn	Fe	Cu	B*
1984-85 av.	2.4	9.3	3.4	33	62	6.8	120	342	1,163	53	79	14	3.0	12	37	0.6	1.2
1991 aver.	3.1	12.5	5.5	30	64	7.2	273	451	1,603	98	126	24	6.8	15	35	1.0	2.0
Differences	0.7	3.2	2.1	-3	2	0.4	153	109	440	45	47	10	3.8	3	-2	0.4	0.8

*Values for K, Mg, Ca, P1, P2, S, Zn, Mn, Fe, Cu and B are in ppm. Figures are averages of tests from several fields.

5. One last note and observation concerning organic matter and CEC. These determine the soil's nutrient-holding capacity, and are usually very hard to increase — a long, slow process. A CEC of four to six means a light sandy soil, with a corn yield potential of maybe 80 to 100 bushels/acre over many years with just sticking seeds in the ground. It takes more fertilizer and water than other soils. This type of soil has a lower nutrient-holding capacity and would have between 0.5 and 1.5 percent organic matter.

If you had a soil like that, you would not expect to change it to the next level, with a CEC of seven to 10. These soils have about 1.5 to 2.5 percent organic matter and a corn yield potential of 100 to 125 bushels/acre with just seed and no fertilizer. They require less fertilizer than sands because they have the capacity to hold more nutrients.

If you had that type of soil, you would not expect to change it to the next level, a CEC between 11 and 13, with organic matter from 2.5 to 3.5 percent and a corn yield potential of 125 to 150 bushels/acre. This soil has little need for additional fertilizer to maintain those yields. A small amount of starter, manure and a good crop rotation should give good results year-in and year-out, with best years getting up to 200-bushel corn (with current knowledge and genetics).

Again, if you had that type of soil, you would not expect to move it up to the next level, with a CEC of 14 to 17 and organic matter of 3.5 to 4.5 percent. The corn yields should range consistently between 175 to 200 bushels/acre with only small inputs: 125 pounds/acre of starter, rotation, manure and/or small amounts of nitrogen.

We could continue further. There are soils with CECs of 20 or 30. These have a high clay content and should have yield potentials of 200 to 250 bushels per acre of corn, although most of this land is not managed to get those yields. Note that you can increase your soil's CEC by increasing organic matter (humus), thereby greatly improving crop yields *plus* reducing off-farm inputs. This doesn't even take into consideration improvements in crop quality and resulting improvements in animal productivity, lower feed supplement needs, and lower animal health care expenses all due to improved soil. Profitability can soar.

A last thought — what can we accomplish when we farm biologically, promoting life in the soil and balancing soil nutrients? You hear so many negative things about farm profits and soil losses. It seems to me that the opportunities with today's knowledge are almost unlimited. Again, where will your farm be in five or ten years?

Raise Your CEC and Your Profits

CEC (cation exchange capacity) is a measure of your soil's ability to store and hold plant nutrients. It is an indicator of the soil's potential fertility. The higher the CEC, the higher the yield potential.

Nutrients are held on the soil's clay and humus particles. Textbooks often assume a certain soil type's CEC is fixed, but many biological farmers have experienced an increase in their soil's CEC, mainly through an increase in humus. The "organic matter" figures on typical soil tests include humus, but they also include raw, undecomposed organic matter, which does not increase CEC. A special humus test will give a better measure.

Let's see how raising your soil's CEC can increase your crop yields and your profits. The table below shows typical corn yields from various CEC soils.

CEC	Average corn yield (bu/acre)
5	80
6	90
7	100
8	110
9	120
10	130
11	140
12	150
13	160
14	170
16	180
18	190
20	200

If you go from CEC nine to CEC 13 (as biological farmers have done), the corn yield potential increases 23 percent, or 30 more bushels per acre. At a price of $2.00/bushel, this is an additional profit of $60.00 per acre. Of course, growing higher yields also requires higher plant populations and more available nutrients throughout the growing season. Good, well-aerated soil structure is vital, and with high biological activity (earthworms, beneficial microbes), still higher yields can be attained as some biological farmers have proved.

The CEC measures how big your soil's "gas tank" is, but if you fail to fill the tank with "fuel" (nutrients from organic matter and fertilizers), your crops can "run out of gas" halfway through the season.

The CEC is correlated with soil type. Clay and loam soils have a larger nutrient storing capacity than sand because of their higher clay content, but increasing any soil's humus level will raise its CEC. Humus also provides many other benefits in the soil including improvement of soil structure, drainage and water retention; reduction of erosion; and it provides plants with improved growth and disease resistance.

10

LIMING & pH

Assuming that you have had your soil tested and have decided that you need to correct some serious imbalances, what do you do first? I can't make any blanket statements or recommendations, simply because soils are very different in different regions. In most soils of the eastern United States, the leaching from higher rainfall results in calcium levels being lower than the "ideal" range of 70 to 85 percent base saturation. In contrast, western U.S. soils often are adequate to high in calcium and are low in other elements. Obviously, your approach to soil balancing should be to raise the levels of low or deficient elements and to lower (or at least not raise) the elements that are in excess. It will probably take several years at least, and some soils may never, in practical terms, be able to be balanced to the "ideal" proportions. Still, moving them toward the ideal should improve crop performance. It may be that concentrating your efforts toward "row support," or providing well-balanced fertility in the row, and not trying to correct all of the soil is the most economical and practical method.

Since low calcium is so often the case in abused agricultural soils, we will begin with this element. In many ways, calcium is the key to building soil fertility. It improves soil structure, thus increasing aeration which is vital for good root growth and the beneficial forms of soil life. Higher levels of soil calcium increase availability of the other plant nutrients by such mechanisms as raising the soil's CEC and buffering capacity, increasing root growth and increasing microbial release of tied-up nutrients.

Calcium is a vital element in plant growth and health. It is necessary for strong cell walls, for cell division, for normal functioning of cellular membranes involved in nutrient uptake and energy release, and in helping prevent invasion of disease pathogens. By increasing root growth and regulating so many cell functions, adequate calcium also improves crop quality. But calcium cannot be transported from one part of a plant to another, so newly formed roots, stems and leaves need additional calcium from the soil. This means that a constant supply of available soil calcium is needed throughout the growing season. Overuse of nitrogen and/or many salt fertilizers leads to acidity or a tie-up of calcium. This can happen in soils that are supposed to have plenty of calcium. Research by Dr. Lloyd Fenn at Texas A&M's El Paso Agricultural Experiment Station found that plants may not get enough readily available calcium, even in high-calcium soils. By supplying a readily available form of calcium along with ammonium-bearing fertilizers, he had yield increases of 53 percent for dry land cotton, 60 percent for lettuce and 56 percent for peanuts, as well as increases in plant hardiness and disease resistance.

Because of the overwhelming importance of adequate calcium, when you are balancing soil and calcium is low, be sure to begin by adding calcium. But what kind of calcium? Most people think "lime" when you mention calcium, and farmers have been conditioned to think of liming as a remedy for soil acidity and use it to correct pH. The subjects of calcium, lime and pH can be confusing and have been used carelessly in the literature on the subject, so let's define and explain them.

What is pH?

The term pH is a chemical abbreviation for the concentration of hydrogen ions, which cause some of the acidity in soil. Technically, pH is the negative logarithm of the hydrogen ion concentration. Therefore, every whole number change in this logarithmic scale is a change of 10 times the hydrogen ion concentration. For example, a pH of 5.8 is ten times more acid than 6.8, and 4.8 is 100 times more acid than 6.8.

Hydrogen ions (H^+) only cause soil acidity as measured by a standard pH test when they are in water solution. This is called *active acidity*. Most hydrogen ions in typical soil are loosely attached (adsorbed) to soil particles, along with other positively charged ions (cations), such as calcium, magnesium, potassium, sodium and aluminum. The soil particles carry many negative electrical charges, which are what attract the positively charged cations (unlike charges attract, and like charges repel). When hydrogen ions are attached to particles, they are called potential acidity, and do not contribute to soil pH (as it is usually measured) at that time, but they are present to contribute to acidity if conditions change. If soil acidity is neutralized by liming materials, both active and potential acidity are neutralized. Aluminum ions can also contribute to soil acidity at acid pHs, but pH only measures hydrogen ions, not aluminum.

Many people believe that soil pH is fairly constant, but the pH can vary widely from one specific place to another (such as in the row versus between rows) or from one month to another. The pH immediately next to a root can be one or two points lower (more acid) than the surrounding soil. Fertilizer applications can cause large changes in pH, both short-term and long-term. High ammonia (or high ammonium-containing fertilizer) use and/or high salt fertilizer use tend to lower soil pH.

The pH measures just the hydrogen ions in solution, but there is more to the story. A water solution contains not only hydrogen ions (H^+), but also hydroxyl ions (OH^-), which cause alkalinity. There is a balance between the two. When there are more hydrogen ions, there are fewer hydroxyl ions and vice versa. Scientists use a pH scale to measure the total range of acidity and alkalinity. The pH scale ranges from 0 (most acid) to 14 (most alkaline), with neutral (pH 7) being midway between:

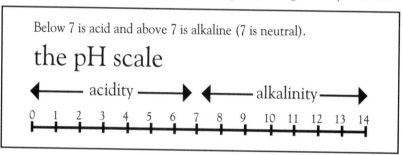

Below 7 is acid and above 7 is alkaline (7 is neutral).

the pH scale

← —— acidity ——→ ← —— alkalinity ——→

0 1 2 3 4 5 6 7 8 9 10 11 12 13 14

The pH of some common materials includes: lemon juice = 2, vinegar = 2.5, black coffee = 5, pure water = 7, sodium bicarbonate solution = 8.2, ammonia water = 11. In pure water, there is an equal number of hydrogen and hydroxyl ions, so the pH is 7, or neutral. If there is something else in the water which either uses up or adds hydrogen or hydroxyl ions, the solution becomes acid or alkaline.

In the soil, there is a complex mixture of components which can affect pH including clay, humus, organic molecules, soluble inorganic salts and carbon dioxide. The soluble salts include carbonates, bicarbonates, nitrates, sulfates and chlorides of calcium, magnesium, potassium, sodium and other minor elements.

A *salt* is made up of two parts, or ions, loosely held together. One ion is positively charged (the cation) and the other is negatively charged (the anion). For example, calcium carbonate is made up of calcium ions and carbonate ions, $Ca^+ + CO_3^-$. Placed in water, some or all of the ions of a salt will separate (dissociate) and become free to recombine with some other ion of opposite charge.

A fertile soil has most of the negative charges on its colloidal clay and humus particles filled with cations which are plant nutrients (calcium, magnesium, potassium, ammonium). As roots absorb nutrient cations and exchange them for hydrogen ions (called base exchange; see Chapter 5), more and more hydrogen ions accumulate, and the soil becomes more acid; that is, the pH drops. Below pH 5, larger numbers of aluminum ions are released from clay particles, also contributing to soil acidity. In general, an acid soil is a low fertility soil. But there are other aspects to soil pH.

Actually, the pH of the soil has little direct effect on plant growth. Experiments have shown that if a plant's roots have access to adequate nutrients and there is no toxicity, moderate to strong acidity or alkalinity causes no problems. But too low or too high pH can adversely affect plant growth in other ways. The availability of the various plant nutrients varies at different pHs, some being less available at low pH (acid) and others less available at high pH (alkaline). Elements such as aluminum, manganese and copper are toxic to plants in too large amounts, so at extreme pHs, plants can suffer from a deficiency of a needed element or from a toxicity of another. (Refer to element availability chart on following page.)

Soil pH also affects the activity of beneficial soil microbes. Bacteria and actinomycetes prefer alkaline pHs, while fungi do best in acid conditions. Plants grow better when beneficial organisms are abundant in the soils (see Chapter 5), so a soil pH that is not too acid and not very alkaline is a good compromise.

For best organism activity and best availability of most nutrients, the soil pH range from about 6.2 to 6.8 is generally recommended as "ideal" for most soils and most crops. However, highly weathered tropical soils which are high in aluminum and iron are naturally very acidic and cannot be raised to over pH 5.5 or 6.0 without degrading soil structure and causing serious imbalances and toxicity from trace elements. Some soils in the southeastern United States are of this type.

Neutralization. Since excessive acidity generally is not good for plant growth and does not allow adequate fertility, farmers will want to eliminate the acidity or neutralize the hydrogen ions or raise the pH ("sweeten" the soil). This is accomplished by adding a so-called liming material to the soil, something that tends to neutralize acidity.

Acidity (hydrogen ions) is neutralized, for example, when a hydroxyl ion combines with a hydrogen ion to form water:

$$H^+ + OH^- = H_2O$$

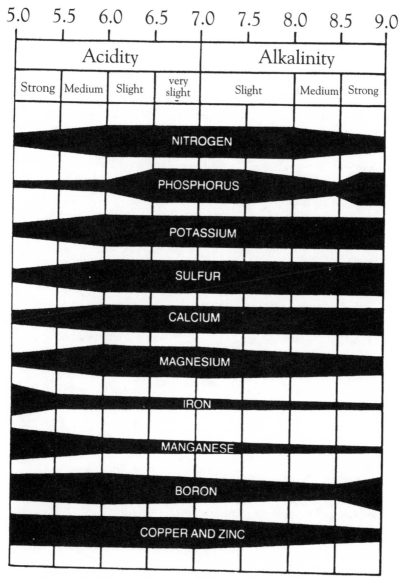

5.0	5.5	6.0	6.5	7.0	7.5	8.0	8.5	9.0
Acidity				Alkalinity				
Strong	Medium	Slight	very slight	Slight		Medium	Strong	

NITROGEN

PHOSPHORUS

POTASSIUM

SULFUR

CALCIUM

MAGNESIUM

IRON

MANGANESE

BORON

COPPER AND ZINC

Availability of Elements to Plants at Different pH Levels for Typical Soils

The most common liming material is limestone, typically calcium carbonate ($CaCO_3$). In the soil, a little of the limestone dissolves in the soil water, releasing the two ions, Ca^{++} and CO_3^-. The carbonate ion combines with two hydrogen ions, neutralizing the acidity:

$$CO_3^- + 2H^+ = H_2O + CO_2$$

In this case, both water and carbon dioxide are by-products. The calcium ion that is left over can then become attached to a soil particle, helping to restore the soil's supply of plant nutrients.

Now, perhaps you can realize a couple of important facts. First, it is not the calcium that neutralizes soil acidity; it is the other ion that is attached to the calcium (in this case the carbonate ion in the limestone). Magnesium carbonate (a part of some limestones called dolomite or dolomitic limestones) is also an effective neutralizer of soil acidity — and it contains no calcium. Also, some calcium-containing materials do not raise soil pH; they might even make the soil more acid. An example is gypsum, calcium sulfate. The sulfate ion could combine with hydrogen to form sulfuric acid, H_2SO_4.

So, once again, calcium *does not* neutralize acidity, but since most commonly used liming materials contain calcium, the idea that calcium neutralizes acidity has become widespread.

Carbonate limes, either calcium carbonate or calcium magnesium carbonate, should only be used if the pH is low. Trying to change the soil Ca:Mg ratio at pH 7 or above using lime is not effective, or is extremely slow. These carbonate limes are low in solubility and soil acidity improves their solubility. So if the soil pH is neutral and you still need to add calcium, the use of gypsum and/or hydrated lime work better. Carbonate limes have two or fewer pounds of soluble calcium per ton, gypsum has 25 pounds and BioCal or hydrated lime with sulfur present have 150 pounds of soluble calcium per ton.

Another thing to realize is that the most important result of liming soil is not neutralizing the acidity, but replacing the cation nutrients that have been depleted by plants. Of these cations, the most important is calcium, since it normally occupies about 70 to 85 percent of the negatively charged sites of soil particles (in fertile soils), with magnesium about 12 to 18 percent and potassium 3 to 5 percent.

Why lime?

The subject of liming soils is complicated by the effect of pH on plant nutrient availability and soil life and by the particular cations that come along with the neutralizing ions. Different crops are said to "prefer" a certain soil pH. For example, alfalfa "prefers" pHs from 6.5 to 7.5, corn and wheat 6.0 to 7.0, but rye likes 5.5 to 6.5. Is it really the pH that these crops "prefer," or is it that some crops, like legumes, need higher levels of calcium than grass crops? Since soils are usually limed with calcium-supplying materials, high-pH soils usually provide more of the needed calcium. You still may need to supplement with a soluble source.

Dr. William A. Albrecht showed the truth of this in a paper entitled "Soil Acidity as Calcium (Fertility) Deficiency," published in 1952 (*Research Bulletin No. 513*, University of Missouri Experiment Station; G.E. Smith co-author). He grew soybeans and spinach in highly acidic soils (pH 4.0, 4.5 and 5.0), but he supplied them with a balance of nutrients including plenty of calcium. Other plants were grown in similar soils, but with less and less calcium, or none at all. The plants grew perfectly well as long as there was sufficient calcium present. The "low-calcium" and "no-calcium" plants were severely stunted. Growth was good when the base saturation of calcium was at least 60 percent, although much better growth resulted above 75 percent calcium. Soybean nitrogen-fixing nodules only developed normally when there was adequate calcium (in fact, Dr. Albrecht found that nodules would not develop in pH 7.0 soil if there was not enough calcium). The test plants took up not only more calcium in high-calcium soil, but more of all nutrients; thus their nutritional value as food was much higher.

Research

Most scientific tests that determine the effects of soil nutrients on crops usually only measure yield, or *quantity*. Rarely are such things as crop diseases, pests, nutritional value or storability measured — things we usually call *quality*. Rarely does a scientific study last long enough to follow the long-term effects of applied fertilizers or liming materials. Many studies are conducted with plants in pots in a greenhouse, and results may or may not apply to field conditions. Sometimes soil is sterilized beforehand, while in real life field soil is populated by many organisms, and certain ones are essential to healthy, high-yielding crops. Agricultural scientists almost never follow through and feed their test crops to animals to see their real nutritional value. Nutritionists do that sort of thing, and they may not care how a crop was fertilized or whether the soil had beneficial organisms.

Yet such total, long-range, full-circle testing is not that difficult or expensive. Decades ago, Dr. William A. Albrecht found that small animals such as rabbits and guinea pigs can adequately substitute for livestock in testing (bio-assaying) crop quality. Dr. Albrecht and his associates found that alfalfa and clover fertilized with plenty of calcium (lime) resulted in more efficient weight gain (more gain on less feed) and better quality wool when fed to sheep, rabbits and guinea pigs. Similar research in Wisconsin using alfalfa and Ladino clover fed to guinea pigs found that moderate, balanced soil fertilization produced nutritionally superior feed (best weight gain), while heavy fertilization decreased feed value.

The amounts of calcium that soil scientists used to think were sufficient are in some cases not enough for quality crops or even average growth. Part of the reason is that calcium does not readily move from one part of a plant to another. It cannot be transferred from older leaves to younger or from leaves to developing grain or fruit. A continuous supply of calcium is needed for peak yield and quality. Measuring soil calcium at the beginning of the growing season or an experiment does not assure that there will be enough throughout the life of the plant. Soil aeration and organisms are important. Suffocated roots cannot grow out into the soil and absorb more calcium, and adequate calcium is needed for root growth. Too high levels of other cations — magnesium and potassium — will decrease the amount of calcium a plant will take up. Calcium *can* be a limiting factor in crop production.

What About Ratios?

A subject that causes some controversy with regard to liming materials is the importance of the relative amounts of soil nutrients, usually expressed as ratios. Based on numerous studies, some soil scientists have written that the ratio of calcium to magnesium is not important (within limits) and that as long as there is enough calcium and magnesium to replace what the crop removes, there are no problems. Further, they say that if dolomitic lime (which contains both calcium and magnesium) is cheaper than calcitic lime (high-calcium lime, calcium carbonate), then it is better to use dolomite. Specifically, they say that various Ca:Mg ratios ranging from about 0.5:1 (high magnesium) to more than 30:1 (high calcium) do not affect the yield of crops. That may be true, but as we mentioned earlier, what about quality, especially food value? And what about possible effects on soil structure? And humus? And soil life? These trials were done using what they call optimum fertilization. What would the results be if you cut the fertilizer in half?

Do ratios of soil nutrients matter? We have covered the fact that plants need some nutrients in relatively large amounts, others in lesser amounts and the trace elements in very small amounts. Obviously, the relative amounts of different elements *do* matter. In fact, some of them, especially some of the trace elements are toxic in too large amounts. A balance is important. But how critical are exact ratios? And who decides which balance is best?

According to a standard soils textbook, the typical proportions of cations in clay soils are 75 to 85 percent calcium (base saturation), 12 to 18 percent magnesium and 1 to 5 percent potassium (L.M. Thompson & F.R. Troeh, *Soils and Soil Fertility*, 4th ed., 1978, p. 167). If we calculate the

Ca:Mg ratios, they would range from about 4:1 to 7:1, with much more calcium than magnesium. But plant absorption of these two elements averages about 1.6:1, with different crop species requiring somewhat different ratios (using textbook figures, alfalfa uses them in a 5.25:1 ratio, while corn uses nearly equal amounts of calcium and magnesium).

If you were to grow only one kind of crop, you could "fine-tune" your soil to that crop's needs, but most growers use a crop rotation of some kind, usually including a legume crop. It makes sense to fertilize soil for the most valuable crop and to rotate crops so that the preceding crop supplies what the next one needs. Usually the legumes in a rotation are high-value crops, perhaps alfalfa, soybeans, or other beans or peas. Legumes need larger amounts of calcium than magnesium. Calcium is important in many cellular functions, and no matter what the crop, higher calcium improves root growth, disease resistance and crop quality. Calcium improves soil structure and stimulates beneficial soil organisms. For those reasons, I like to see calcium-to-magnesium ratios of at least 5:1, or higher.

Of course, plants do need magnesium, too. It is essential for protein synthesis, chlorophyll production and as an activator of many cell enzymes. Some soils are definitely short of magnesium, especially leached and sandy soils with a low CEC.

There is an interaction between calcium and magnesium — and potassium as well. High soil potassium decreases the plant's uptake of both calcium and magnesium. Low magnesium and calcium, and high potassium in feeds leads to livestock health problems, such as grass tetany symptoms (downer syndrome, displaced abomasum, to mention two). High soil magnesium, as usually occurs in areas with dolomite bedrock, can lead to lower quality legume crops.

Some scientists worry that excessive use of high-calcium lime will result in too low magnesium levels in crops, which certainly can happen. That is why we like to strive toward a certain balance and not overdo a good thing.

Liming Materials

Commonly used liming materials include:

1. *High-calcium lime* (calcium carbonate, calcitic limestone, calcite, aragonite, $CaCO_3$) — with 32 to 40 percent calcium and less than three percent magnesium. Has fairly low solubility (slow acting); best incorporated into upper soil.

2. *Dolomitic lime* (calcium magnesium carbonate, dolomite, $CaCO_3 \cdot MgCO_3$) — variable, with about 22 percent calcium and eight to

20 percent magnesium. Low solubility; best incorporated into upper soil. Not always a good plant source of calcium.

3. *Marl* (calcium carbonate plus impurities) — variable percent of elements; an impure form of high-calcium lime. Other high-calcium materials include oyster shell, chalk, paper mill sludge, sugar-beet waste, and water treatment by-product. These have limited usage, and some may have too high levels of toxic heavy metals.

4. *Liquid lime* (fluid lime; a suspension of very fine particles of any liming material) — variable sources and calcium:magnesium contents. Good plant availability, but expensive because of water content.

5. *Quicklime* (calcium oxide, burned lime, unslaked lime, CaO) — 60 to 71 percent calcium. Very caustic and difficult to handle; highly reactive in soil; seldom used.

6. *Slaked lime* (calcium hydroxide, hydrated lime, CaOH) — 45 to 55 percent calcium. Caustic and difficult to handle; highly reactive in soil.

7. *Kiln dust* (calcium oxide and calcium hydroxide plus other elements) — 28 to 36 percent calcium, up to 5 percent potassium, 2 to 4 percent sulfur. A by-product of cement or burnt lime manufacture. Good solubility and plant availability; can be surface applied; difficult to handle.

8. *Basic slag*, blast furnace slag (variable composition; mixtures of calcium silicate, calcium oxide and calcium hydroxide; may contain magnesium and phosphorus and other elements) — about 29 to 32 percent calcium. A by-product of steel manufacture.

9. *Fly ash* (variable composition; mixture of calcium oxide, calcium hydroxide and calcium carbonate, plus sulfur, boron and molybdenum). Fine particles; difficult to spread.

The speed and effectiveness of these liming materials depends not only on their composition and solubility, but also on the size of their particles. Materials with lower solubilities, such as limestones, should be finely ground or applied far ahead of their desired action. In fact, dolomitic lime is such a hard stone that it can take surface-applied dolomite 10 to 13 years to neutralize acidity two inches below the surface. Consultants recommend applying and incorporating it about 18 months ahead of time. High-calcium lime acts somewhat faster at first, while the very fine particle size of such materials as liquid lime, kiln dust and fly ash give fast, same-year action.

For soils that are low in magnesium, most people recommend dolomitic lime, but again, its slow action is a disadvantage. Better, quickly available magnesium sources include sul-po-mag (potassium magnesium sulfate, sulfate of potash-magnesia, or K-mag) and magnesium sulfate (epsom salts); however, neither of these materials is a liming material (they do not neutralize acid). Some research has found that dolomite does not supply enough calcium for some crops (corn, potatoes and other vegetables).

If you want a calcium-supplying material but do not want to raise the soil's pH, as on alkaline soils, then an excellent material to use is gypsum (calcium sulfate). It does not neutralize acid and contains 22 to 23 percent calcium, plus 18 percent sulfur, a valuable nutrient. Another excellent source is kiln dust with sulfur. It is higher in solubility than gypsum, and because of the fine grind, can be surface-applied. It gives excellent results the first year. It works more like a fertilizer than a liming material.

The following table gives the results of a five-year comparison of high-calcium lime, gypsum and BioCal (a blended, formulated kiln dust product from Midwestern Bio-Ag). They were the only soil correctives on the fields, but a balanced crop fertilizer was used. The field has been on a corn/soybean rotation with rye following soybeans and the corn interseeded with rye grass and clover. No herbicides or insecticides were used. Yields have been good to excellent. Corn yields ranged from 140 to 240 bushels per acre. Soybeans have been in the 50 to 65 bushels per acre range. The only tillage besides cultivation is a shallow incorporation of residues with a Howard Rotavator.

1991 field		High-cal lime 1 ton/acre		Gypsum 1,000 lbs/acre		BioCal 1,000 lbs/acre	
Average:		1995	1997	1995	1997	1995	1997
O.M.	2.1 L	1.9	1.9	2.8	2.2	2.5	2.3
ppm P1	89 VH	75 VH	87 VH	82 VH	75 VH	127 VH	105 VH
ppm P2	141 VH	114 VH	153 VH	150 VH	131 VH	217 VH	153 VH
% base saturation:							
K	5.2 VH	4.6 VH	4.5 VH	3.8 H	3.9 H	5.1 VH	4.5 VH
Mg	30.4 VH	33.7 VH	31.7 VH	28.8 VH	28.9 VH	29.4 VH	27.6 H
Ca	64.4 M	61.7 M	63.8 M	67.4 M	67.2 M	65.5 M	67.9 M
pH	7.3	7.0	7.3	7.3	7.2	6.9	7.4
ppm:							
S	21 H	10 L	28 H	16 M	25 H	12 L	29 H
Zn	2.0 M	2.8 M	3.1 M	4.4 H	4.3 H	4.2 H	4.1 H
Mn	16 H	17 H	17 H	17 H	13 M	16 H	13 M
Fe	19 H	51 VH	34 VH	47 VH	33 VH	43 VH	35 VH
Cu	0.6 L	1.0 M	1.2 M	1.0 M	1.1 M	1.3 H	1.3 H
B	1.2 M	0.7 L	1.0 L	0.7 L	1.2 M	1.0 M	1.4 M

(soil tested by Midwest Lab, Omaha)
L = low, M = medium, H = high, VH = very high

Comments:

First of all, the soil test numbers are guides and give trends. This field was a high P/K testing field when we started. CECs range from 10 to 12, and it is a well-drained silt/loam. No matter which lab you send the samples to, they should come back high in P and K. The numbers may be different, but results the same. The soils in Wisconsin are high in magnesium and medium to low in calcium. Except for some trace elements, the only soil correction would be adding more calcium and less magnesium. Because the pH is near or above neutral, lime would not be the preferred source of calcium. Gypsum and/or BioCal fit better. They provide a more soluble calcium plus sulfur, which should also help to reduce magnesium if drainage is good.

For whatever reason, the organic matter levels increased on the gypsum and BioCal fields, but not on the lime field. As for phosphorus, the levels are still in the VH range, but the BioCal numbers trend higher, with no added phosphorus. Potassium levels have held well. There appears to be a trend toward lower K levels in the gypsum plot. The row fertilizers for corn and soybeans are:

	Corn row fertilizer 350 pounds/acre	Soybean row fertilizer 200 pounds/acre
N	15%	12%
P_2O_5	8%	9%
K_2O	2%	3%
Ca	3%	8%
S	16%	13%
B	0.12%	0.15%
Cu	0.12%	0.15%
Mn	0.4%	0.45%
Zn	0.5%	0.6%

Ingredients of Row Fertilizers:

A blend of Hartland ammonium sulfate, MAP (monoammonium phosphate), sulfate of potash, North Carolina rock phosphate, Idaho soft rock phosphate, calcium sulfate, granulated calcitic lime, sea kelp, fish meal, molasses, diatomaceous earth, manganese sulfate, copper sulfate, calcium borate, zinc sulfate. This is a blended, homogenized mix with a final pH of 5.5.

Note that because soil K levels are very high, we use a row fertilizer that is low in K. These soils have maintained K levels with no corrective, crop or maintenance K. Soil testing is a way of monitoring what is happening. If K levels were dropping, we would change row fertilizer and add more, or bulk-spread a soil corrective. We would also question our tillage, green manure crop program and evaluate earthworms and biological activity. This program should sustain our soil fertility or else we will be growing crops by robbing from the soil's reserve, and our future would look bleak. Remember, we test to give direction and clues, and to monitor our program.

Now look at magnesium:calcium ratios. Adding the high-calcium lime doesn't seem to change anything. We believe this is due to its low solubility. If the soils were acid, it should work better. Both gypsum and BioCal are improving these ratios. A recent test has the BioCal field at 25 percent Mg and calcium over 70 percent. What about quality and yields? Quality is hard to measure without feeding trials. The yields have been slightly higher on the BioCal and gypsum plots. The future is exciting as these soils improve; our yields can improve or we can get the same yields with lower input. As for quality, in the feeding chapter, I will show what is happening with this type of fertilizer program. How much value do we put on the increase in OM?

Now look at trace elements. We have not bulk spread any correctives. As for sulfur, note from the row fertilizer that we are getting 40 to 50 pounds per acre, plus on the gypsum plot, another 170 pounds (gypsum contains 17 percent sulfur), and on the BioCal plot, an additional 60 pounds. This is not reflected in the soil test. Sulfur is an anion and will leach out provided we have no hardpan. The numbers on the test jump all over based on rainfall and time of application. The theory is that the extra sulfur binds with magnesium, forming epsom salts (magnesium sulfate) and leaches out. It appears that this is happening.

Zinc, copper and boron levels tend to be improved in the gypsum and BioCal plots over the lime plot. All plots get the same application rate. This could be due to the higher organic matter levels, better biological activity, improved nutrient cycling or any other logical reason you could come up with.

As for manganese and iron, these levels move based on tillage (more oxidation), use of harsh fertilizers and biological activity. Manganese soil tests give you clues to the activity at the time of testing. If the pHs are neutral or above, and you have high levels of manganese, the biological activity at that moment is good. We like a ratio of 1:1 for Mn:Fe. This means oxidation and biological activity are in balance as they should be. Again, these are more clues as to what could be happening in the soil.

11

FERTILIZERS: WHAT YOU GET

Fertilizers are sources of plant nutrients that the farmer adds to the soil to supplement its natural fertility. Soils vary in their natural fertility — their cation exchange capacity, pH, organic matter content, the stored mineral elements and their balance.

Under ideal conditions a fertile, "living," aerated soil can supply all the nutrients a growing crop needs, but ideal conditions rarely exist. Most soils have some imbalance or deficiency of nutrients. Changing weather conditions also affect nutrient release and availability. A soil and crop usually need a supplementary fertilizer, especially with today's high plant populations and high yield expectations.

When crops are harvested, part of the nutrients are removed, and unless crop residues or manure are recycled back into the soil, its supply of available nutrients can be depleted. Recall from Chapter 5 that soils have large amounts of reserve nutrients, but they are mostly tied up in minerals and organic matter and are not immediately available to plants. A small portion of tied-up nutrients can become available by the action of roots and soil microorganisms.

About 95 percent of a crop's growth actually comes from the air and water (carbon, hydrogen, oxygen), and only about 5 percent comes from soil nutrients. Still, soil nutrients are vitally necessary for normal growth, crop health and quality. They are needed in the proper amounts, in the right balance and at the right time.

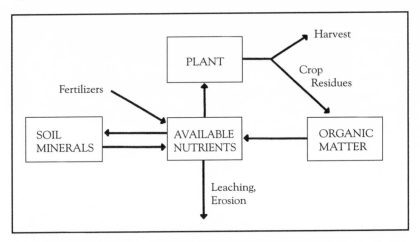

The life of the soil — the bacteria, fungi and earthworms — plays a major role in nourishing a crop. They break down and recycle organic matter. They modify the soil pH and bring it toward the "ideal" range of 6.2 to 6.8. They help break down the minerals of the soil and release tied-up phosphorus, potassium, calcium, magnesium and others. They hold nutrients in a chelated form, which plants can readily absorb. They feed nutrients to plants slowly over the growing season, at the rate plants need them. Some fungi (mycorrhizae) live in plant roots and actually aid roots to absorb additional water and nutrients — more than they could absorb otherwise. Nitrogen-fixing bacteria provide additional nitrogen from the air. Many microbes that live on roots protect them from invading diseases. These are some of the reasons biological farmers emphasize the life in the soil and do everything possible to help it flourish.

History of Fertilization

Early in history, man discovered that poor soil could be enriched by adding such things as limestone, ashes, manure, fish or seaweed. This is called *amending* the soil. These early fertilizers contained several elements, including trace elements — and often organic matter. In the 17th, 18th and 19th centuries, chemists and other scientists discovered that plants will grow when various chemical salts such as potassium nitrate are added to the soil. Early commercial fertilizers were made by mixing bird guano and potash salts.

The commercial fertilizer industry has grown rapidly in the 20th century, especially since World War II. In the last 50 years, commercial fertilizers have tended to be pure salts, primarily N-P-K (nitrogen, phosphorus, potassium), with few or no trace elements. With intensive agriculture and crop removal, many soils have become deficient or unbalanced in several elements (in plant available form), including trace elements.

Still, the widespread use of commercial fertilizers has greatly increased the world's food supply, reducing hunger and making possible a high standard of living. The fertilizer industry and research scientists have overemphasized the importance of N-P-K and neglected the other elements, including a proper balance of *all* elements. Out-of-balance soils and unhealthy, poor quality crops are the result, even though yields (quantity) may be high. Feeding low quality food to animals and people often results in poor health and high medical expenses. Soil structure, humus and beneficial organisms have been neglected and some of the concentrated, high-salt commercial fertilizers even contribute to soil degradation.

How Much?

In typical farming situations, adding a fertilizer that provides elements which are low in the soil does give a crop response, and sometimes a dramatic improvement. Adding more fertilizer gives still better crops, but only up to a point of diminishing returns, both in the crop response and in economic profitability.

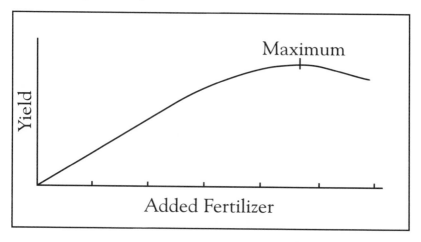

Some crops and some varieties give a greater response to fertilization than others. Small grains need lower levels of nutrients than vegetables. Legumes need little or no added nitrogen because they get nitrogen from their root nodules. Corn gives a large response to nitrogen. The soil's ability to hold nutrients (its cation exchange capacity) also determines the level of fertilization that a person might use.

Terms

There are several terms used to describe fertilizers that can be confused and misused:

1. *Organic* can mean a material produced by or derived from living things, such as animal manures, green manures, fish emulsion and crop residues. Organic materials tend to break down slowly, providing slow release of nutrients. But the time lag may be too long, and if the soil is cool or biological activity is low, there may be little breakdown at all.

Another definition of organic, used by chemists, is any compound that contains carbon, whether or not it is natural or man-made. Most life-produced materials (the first definition) contain carbon, but such synthetic compounds as many herbicides and insecticides are called "organic" (second definition). Urea is a fertilizer material that is organic and could either come from natural sources (animal urine) or be man-made.

2. *Inorganic* simply means not organic. It includes both natural (occurring in nature) and synthetic (man-made) materials. The term chemical is sometimes used loosely to refer to synthetic fertilizers or pesticides. Technically, all matter is made up of chemical elements, whether natural or synthetic. Also, some salt fertilizer materials, such as potassium chloride, are found as natural rock deposits, so are not synthetic.

3. *Salt* refers to a chemical compound made up of a cation (positively charged ion) and an anion (negatively charged ion), such as common table salt, sodium chloride ($NaCl$), or fertilizer salts like potassium chloride (KCl), potassium sulfate (K_2SO_4), or diammonium phosphate $(NH_4)2HPO_4$. When salts dissolve in water, the two ions can separate. If they are not used immediately by plants, they can either be held by soil particles (as ammonium, phosphate and potassium are) or be leached downward (nitrate and sulfate). Salt fertilizers usually give quick release of soluble, readily available nutrients — a "shot-in-the-arm" to plants. Too much at once may cause an imbalance, however. Also, some soluble salt ions, such as phosphate, rapidly become insoluble in the soil (and unavailable to roots, unless released by microbes). One of the two ions of some fertilizers, especially the chloride of potassium chloride, can harm crops that are chloride sensitive and also kill soil life. Some salt fertilizers cause greater osmotic damage (drying out effect) to roots and organisms than others. This is measured by the salt index. A higher salt index indicates greater potential crop damage.

Material	Salt index
Ammonium nitrate	104.7
Ammonium sulfate	69.0
Calcium nitrate	52.5
Calcium sulfate (gypsum)	8.1
Diammonium phosphate	34.2
Monoammonium phosphate	29.9
Potassium chloride	116.3
Potassium nitrate	73.6
Potassium sulfate	46.1
Sulfate of potash-magnesia (sul-po-mag)	43.2
Superphosphate	7.8
Triple superphosphate	10.1
Urea	75.4

4. *Form.* Fertilizers can be solid (powders, granular, pelleted, prilled) or liquids (dissolved in water) or a gas (anhydrous ammonia). They can be single compounds (such as ammonia or potassium sulfate) or a blend of two or more materials.

5. *Analysis* refers to the amount of nutrients contained in a fertilizer. The grade is the minimum guaranteed percentage (by weight) of total nitrogen, available phosphorus (as P_2O_5) and water-soluble potassium (as K_2O). A 5-10-10 fertilizer contains at least 5 percent N, 10 percent P_2O_5, and 10 percent K_2O. The same fertilizer could also be said to have a ratio of 1-2-2. The percent of other elements could be added to the N-P-K grade; for example, a 3-15-15-9S would have 3 percent nitrogen, 15 percent P_2O_5, 15 percent K_2O and 9 percent sulfur. A complete fertilizer (as the term is used in the industry) contains nitrogen, phosphorus and potassium. To complicate matters even more, phosphorus (P) and potassium (K) guarantees in the fertilizer industry are based on soluble phosphate (P_2O_5) and potash (K_2O) levels. To convert P_2O_5 to P, divide by 2.29. To convert K_2O to K, divide by 1.4. Since soil fertility is not mathematical, this is a moot point, but many farmers do get confused.

6. *Natural base* means a natural-mined material, not man-made, containing more than the N, P or K units for which it is bought. Because it comes from the earth, it also contains trace elements and possibly other compounds that help plants grow. It is generally a more slow-release material.

7. *Plant food* refers to a fertilizer in a form that is usable by plants, a high quality, non-toxic material. It can be organic, natural or synthetic. Plant food can be applied directly on the seed or sprayed on the plant. It is not harsh. Also, feed grade livestock materials can be used as plant food.

Getting the Most From Fertilizer Dollars

All fertilizer materials are not the same. Research says N is N, P is P and K is K, and it takes X pounds to grow the crop. That may be true. But, if you use compost, green and/or livestock manures, and a combination of natural-mined materials and high quality, balanced manufactured ones to provide those pounds, in just a few years you will see tremendous soil property changes.

It's not just adding so many pounds. It is more complex than that. The secret to making fertilizers work is the balance, the concentration (putting more nutrients in the root zone) and the recovery (that means large root systems with lots of soil life).

In choosing fertilizers, use materials which do not provide extra unneeded components, as KCl (muriate of potash) does. It is a salt and the excess chloride is a problem. Research says it leaches out. That's true, except that it takes time. It doesn't benefit soil life, and we are convinced that when it goes it takes calcium with it. So this cheap potassium source actually may be expensive in the long run.

Then there is the *solubility*. Many commercial fertilizers are too soluble, allowing them to leach before the plant can use them and causing an immediate imbalance for the plant. They can also interfere with root growth, soil microorganism balance and soil microorganism populations.

When we build fertilizers, we choose sources that are moderately soluble, that have some slow-release materials for later use, and that have a balance of all the nutrients. In addition, we add things like humates, kelp, fish, carbon and mined rocks that contain substances not yet called essential, but which I feel are necessary for healthy soil and crops. Since most soils are neutral or higher in pH, we granulate these blends in a homogenized mix and keep the pH low. That means each fertilizer pellet is balanced and the low pH keeps the nutrients available longer in the growing season.

Fertilizer Application

The farmer wants to make efficient use of fertilizers so as to maximize crop response and minimize input expenses. The method of fertilizer application will depend on several factors, including needs of the crop species, soil test levels, soil type and cation exchange capacity, crop rotation sequence, spring or fall application, type of fertilizer (salt index, speed of release, etc.), budget and fertilizer program (crop fertilizer versus corrective, etc.). Methods of application include:

Broadcast — fertilizer is applied uniformly over the field. This is suitable for a soil corrective program, but is relatively expensive per acre for crop fertilizers. The following methods are more economical and tend to feed one season's crop:

In-row — fertilizer is placed in the furrow at planting, possibly below or to the side, rather than on the seed.

Banded — fertilizer is applied in a band to one or both sides of seed or plants.

Top-dressed, side-dressed — fertilizer is applied to the crop after emergence. Top-dressed refers to broadcasting on small grains or forage; side-dressed is banding on row crops.

Foliar — a liquid is sprayed on growing plants, with the nutrients entering through the plant's leaves and stems.

Fertilizing Options

There are different approaches one could use in fertilizing crops. Often, the approach that appears to save money in the short term ends up costing more in the long term. Let's look at them and the common practices followed:

Option 1 — use the cheapest fertilizers.

Practices:

1. Apply a minimum of 200 pounds/acre of potassium.
2. Apply 100 pounds/acre of phosphorus.
3. Apply three tons of local quarry lime every three to five years.
4. Apply 200 pounds/acre of nitrogen for corn.

(A soil test might vary the recommended amounts applied, but crop removal amounts are still added. The results are the same.)

Advantages:

1. Easy — can often get good yields, but may degrade soil, "burn" up humus and cripple soil life.
2. Works well on newly farmed soils and with cheap fertilizers, high grain prices and available cheap money (the way it used to be).

Disadvantages:

1. Profit is not always there.
2. It takes a lot of cash.
3. There is no hope for things to get better in the long run. It is not sustainable.

4. It leads to problems:
    a. Weeds, pests.
    b. Trace element deficiencies.
    c. Erosion and loss of humus; harder soil.
    d. Stress effects are more dramatic.
    e. Diseases, both in crops and animals.
    f. Environmental pollution from excessive fertilizer and chemical use.
5. Produces poor quality feed.
6. Leads to soil imbalance. Things will get worse in time.
7. Can lead to higher chemical dependence by reducing biological activity.

Option 2 — Go to a complete liquid plant food, but do only a minimum to build up and balance the soil.

Practices:
1. Get equipment for liquid application.
2. Soil and plant tissue testing are a must.
3. When and how elements are applied is more critical because fewer units are used.
    a. Seed-applied at planting.
    b. Foliar feed in early morning or late day when plants are most receptive.
    c. The stage of plant growth is also important.

Advantages:
1. Less material handling.
2. An efficient way to apply trace elements and phosphorus.

Disadvantages:
1. Can be very costly; soil-applied nutrients are much cheaper.
2. Still need calcium, potassium, nitrogen and sulfur to be added to the soil.
3. Deficiencies can occur over a long period of time because such a small amount of plant food is added.
4. Must be applied repeatedly during the growing season, adding to labor and fuel costs.
5. Requires special equipment and storage facilities.

Option 3 — the most logical approach.

Practices:
1. Use a complete soil test as a guide.

2. Use a blended "natural base" fertilizer that meets your farm's needs.

3. Balance the soil with natural base, high quality fertilizers that are not harmful to roots or soil organisms.

4. Supply all needed nutrients, along with plant growth stimulants, in a constant, balanced flow.

5. Build the soil with grasses, legumes and manure. Add organic matter, which is food for soil life.

6. Get rapid decay of organic matter so nutrients are released for the next crop. Microorganisms eat first — plants get the leftovers and by-products.

Advantages:

1. Stimulates soil organisms. They improve soil structure (aeration, drainage), release soil nutrients and fight diseases and pests.

2. Stimulates plant root growth. Larger roots mean less fertilizer is needed for the same yield, and you get better crop quality. There is less stress from environmental conditions. A larger root system can absorb more soil nutrients and water (better drought resistance).

3. This option deals with all aspects of the soil:
   a. Physical
   b. Chemical
   c. Biological

4. It works toward improvement. Things will get better with time.
   a. Less input cost, higher profitability.
   b. Improved animal health, production and reproduction.

Disadvantages:

1. Requires a better understanding of soil fertility, soil organisms and management practices.

The Fertilizer Industry — a Comparison

In order to grow high quality, high yielding crops without harming the soil or the environment, and to be profitable in the long run, a farmer needs to be aware of exactly what fertilizer materials he is putting on his soil and what effects they have — on the crop and on the fragile "soil ecosystem." Some fertilizers are overpriced for what you get, and they may contain materials that can harm crops or the soil. Does it make a difference?

Here are some questions to ask before purchasing any fertilizer:

1. Do I need the elements I am buying? (Here is where soil testing comes in.)

2. Are N-P-K my limiting factors?

3. Do I correct soil nutrient excess, or just deficiencies?

4. How do I decide how much to put on?

5. Do I put on what the crop takes off? If so, all 13 known elements or just N-P-K?

6. If my soil test levels are high, do I need anything?

7. "Balance the soil — feed the crop." Do I do both?

8. Is this a "numbers game"? If my soil tests 100 pounds/acre low and I add 100 pounds, is it corrected?

9. What are all the possible ways to increase fertility levels in the soil?

10. Are the materials I'm using beneficial to soil life and plant roots?

11. The fertilizer may be soluble (go into solution), but is it available to the plant one week after application? One month? Six months?

The fertilizer "numbers game" (using potassium as an example)

The total amount of a nutrient in the soil reserve, the amount that is available to the plant, the amount that shows up on a soil test report, the amount in a ton of fertilizer and the amount actually used by a plant are all quite different numbers.

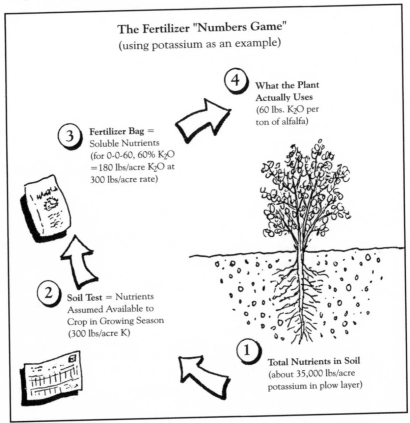

The Fertilizer "Numbers Game"
(using potassium as an example)

4 What the Plant Actually Uses (60 lbs. K_2O per ton of alfalfa)

3 Fertilizer Bag = Soluble Nutrients (for 0-0-60, 60% K_2O = 180 lbs/acre K_2O at 300 lbs/acre rate)

2 Soil Test = Nutrients Assumed Available to Crop in Growing Season (300 lbs/acre K)

1 Total Nutrients in Soil (about 35,000 lbs/acre potassium in plow layer)

The secret to biological farming is to convert a continuous supply of nutrients from the soil (1); to create plant-available nutrients measured through soil tests (2); to minimize purchased inputs (3); and ultimately achieve a point where the plant gets most of its nutritional needs from the soil, therefore requiring only small amounts of supplemental balanced crop fertilizers (4).

12

FERTILIZERS: THE GOOD, THE BAD & THE COSTLY

Plant nutrients are not supplied in their pure, elemental form such as just nitrogen, just phosphorus, just potassium or just calcium. Instead, the nutrient element is combined with one or more elements to form a compound, such as ammonium nitrate, monoammonium phosphate, potassium sulfate or calcium carbonate.

In the soil when water is present, the compound molecule splits apart into two ions, one or both of which are the plant nutrients you want. The other ion may or may not be helpful. For example, consider two potassium fertilizers. The sulfate ion of potassium sulfate provides sulfur and helps make other nutrients more available. But the chloride ion of potassium chloride is excessive beyond plant needs and is harmful to some plants and soil organisms.

When deciding what fertilizers to use, you should be aware of their *total composition*, not just the nutrient elements you want. Also, knowing something about the fertilizer's properties and its long-term effects on plants, soil and soil life will help you make the best choice. Different fertilizers also vary in availability (how easily the plant can use it), so it is not just a number on the bag. Sometimes the "cheapest" source of a nutrient is no bargain. It may cost you in lower crop quality, more pests and weeds, poorer soil structure and worsened animal health and production.

We will now take a close look at the most commonly used materials to supply the various plant nutrients, with comments on their suitability for biological agriculture.

Nitrogen

Plants need more nitrogen than any other element obtained from the soil and it is often deficient during the growing season. Nitrogen is used in many plant functions; it is part of all proteins, enzymes, DNA and many other metabolic molecules. Nitrogen can move around in the plant (it is mobile), and proportionately more is needed in early growth stages. Symptoms of nitrogen deficiency include slow growth of tops and roots, with older leaves turning yellow then brown, especially near the center.

Nitrogen fertilizer cost can be reduced by growing legume crops in rotation, by increasing organic matter (green manures, animal manures) and by side dressing a split application on crops such as corn. Broadcasting nitrogen ("weed and feed") and using anhydrous ammonia are wasteful ways of applying nitrogen. The most efficient use of nitrogen is applying some in the starter fertilizer, followed by a side dressing of a dry or liquid in the root zone at cultivation. The amount to apply will depend on the type of soil, biological activity, applied manure, rotation and desired yield.

Nitrification inhibitors are sometimes used to slow the rate of bacterial change of ammonium nitrogen into nitrate nitrogen. They act by inhibiting or killing the nitrifying bacteria. The bacteria later recover. I prefer not to interfere with natural systems, but loss of nitrogen can be accelerated in hot weather, so this is one way to retain it as non-leachable ammonium. A better way is to frequently recycle organic matter into the soil and to get nitrogen from legume crops and nitrogen-fixing soil organisms. Put the nitrogen on when and where the crop needs it.

Ultimately, all fertilizer nitrogen comes from the air. Nitrogen gas makes up 78 percent of the atmosphere, but plants cannot use pure nitrogen. Instead, they use mainly nitrate (NO_3^-) and ammonium (NH_4^+). Ammonium is a cation (positively charged) and can be held by soil particles, but nitrate is an anion (negatively charged) and is subject to leaching.

The earth's nitrogen is constantly changing from one form to another and moving from the air to the soil, plants and animals, in the nitrogen cycle:

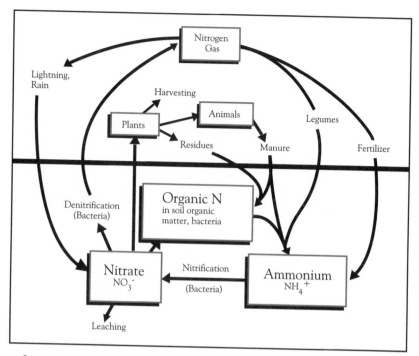

Important processes in the nitrogen cycle include, (1) fixation of atmospheric nitrogen by lightning and microorganisms into ammonia and nitrate; (2) temporary immobilization in proteins and other organic molecules; (3) mineralization of organic matter nitrogen back into ammonia and nitrate; (4) soil bacteria change ammonia into ammonium (called ammonification), ammonium into nitrite and then nitrite into nitrate (called nitrification); (5) other bacteria can change nitrate into gaseous nitrogen and nitrous oxide (the process is called denitrification), which can escape into the air. Other losses of soil nitrogen from the root zone occur from leaching of nitrate, soil erosion, escape of ammonia gas and removal by harvesting.

Nitrogen Fertilizer Sources:

1. Ammonium nitrate (33-0-0). Dry = 33 percent nitrogen. Liquid = 28 or 32 percent nitrogen.

   a. Comments and uses: Contains two forms of nitrogen, nitrate and ammonium. Nitrate is easily used by plants but leaches easily; ammonium stays longer in the soil, but is eventually converted to nitrate by soil bacteria. An excellent corn side dress. Liquid 28 percent works well with contact herbicides when mixed with humic acid, allowing much less herbicide to be used.

b. Limitations and recommended amounts: Fall application on corn stalks for rapid decay: 10 gallons. Contact herbicide use: two to three gallons. Side dress: 10 to 30 gallons (dry: 200 pounds/acre).

2. Ammonium sulfate (21-0-0-24S) — 21 percent nitrogen, 24 percent sulfur.
  a. Comments and uses: Excellent for fall or early spring application on corn, small grains and alfalfa. Has a warming effect on soil in the spring; extends the growing season. Stabilizes soil temperature for improved biological activity. Provides much-needed sulfur in sulfate form (available to plants). Excellent in starter fertilizers. The black form of ammonium sulfate is superior, since it contains carbon. It is also easier to handle.
  b. Limitations and recommended amounts: Ammonium sulfate can only provide part of the crop's nitrogen needs, due to its approximately 1:1 nitrogen to sulfur ratio (a 10:1 ratio is desired). For corn: 200 pounds/acre. For alfalfa: 100 to 150 pounds/acre applied early in the spring.

3. Urea (46-0-0) — 46 percent nitrogen (dry).
  a. Comments and uses: Has only one form of nitrogen (ammonium). Can release high amounts of nitrogen as ammonia gas (toxic to roots and soil life). Best use is broadcast and worked into soil right away (it will release ammonia into the air). Many foliar plant foods contain urea in small amounts. Foliars do work well. If using it as a side dress, keep it away (at least six inches) from the seed and below it. Urea is not my preferred form of nitrogen.

4. Anhydrous ammonia (82-0-0) — 82 percent nitrogen.
  a. Comments and uses: Very highly concentrated ammonia, a toxic gas. Not an effective nitrogen source: 200 units of anhydrous ammonia will perform no better than 80 units of split-application nitrogen. "Least expensive," but actually the most costly form, because:
    1. It's too concentrated. Sandy soils cannot handle high rates; it can be lost into the air as gas.
    2. Very hard on soil life: kills bacteria, fungi, and earthworms near the injection zone.
    3. Causes drastic changes in soil pH and temperature at time of injection, and dries out the soil near the injection zone.
    4. Makes humus soluble, vulnerable to leaching. Oxidizes humus.

5. Can lead to harder soil, from changes in soil structure and loss of humus.

6. Causes long-term increase in soil acidity.

7. Burns and kills roots if applied too close.

8. Causes accelerated plant growth — plant top gets ahead of roots. Should not be used in a biological farming system.

5. Other commercial nitrogen sources:

a. Potassium nitrate (13-0-44) — 13 percent nitrogen, 44 percent potash.

b. Calcium nitrate (15-0-0) — 15 percent nitrogen, 21 percent calcium.

c. Fish emulsion (liquid fish) — 4 to 11 percent nitrogen, plus many other nutrients.

Note: These three sources are expensive per unit of fertilizer, but when used in foliar sprays, do an excellent job of providing high quality nutrients.

d. Liquid nitrogen mixes — 28 percent, 32 percent nitrogen, or others. Made by dissolving ammonium nitrate and urea in water (28 percent), and sometimes ammonia also (32 percent). Good as liquid side dresses.

e. Aqua ammonia — 20 to 24 percent nitrogen. Made by bubbling anhydrous ammonia through water. A good source of liquid nitrogen, but can burn roots and tops if applied in too large amounts.

f. Ammonium thiosulfate (12-0-0-26S) — 12 percent nitrogen, 26 percent sulfur. A 60 percent water solution; can be directly soil-applied or used in irrigation systems. It is compatible with liquid nitrogen and most N-P-K solutions.

g. Monoammonium phosphate (11-52-0) "MAP" — 11 percent nitrogen, 52 percent phosphate. Excellent in starters; does not burn roots.

h. Diammonium phosphate (18-46-0) "DAP" — 18 percent nitrogen, 46 percent phosphate. Commonly used in commercial blends, but too high pH, releases ammonia gas (toxic) and can burn roots.

6. Good organic nitrogen sources:

a. Compost — contains an average of 2 percent nitrogen. Very effective because it is a balanced plant food. It also has plant growth stimulators, and humus to improve soil structure.

b. Animal manures. Poultry manure is highest in nitrogen (up to 8 percent). Manure is excellent in a shallow application in the fall so soil organisms can convert it to forms usable by the plant. Too high amounts are bad and can kill or upset the balance of soil life (see Chapter 14).

c. Green manures (plow downs) — provide 0.5 to 5 percent nitrogen. Legumes such as alfalfa, sweet clover, red clover, beans, etc. provide the most nitrogen. Grasses (rye, etc.) are also good in the growing stage. Green manures are valuable for more than nitrogen; they build soil structure, hold moisture and feed soil organisms.

Phosphorus

Although plants take up less phosphorus than they do nitrogen, potassium and sometimes even magnesium or calcium, phosphorus is often a limiting factor in crop production. It is considered a major element because of its vital need by plants. Adequate phosphorus gives good root systems, early maturation and crop quality. Phosphorus can move from one part of a plant to another, and it is needed throughout the lifetime of the plant. Symptoms of phosphorus deficiency include slow growth, with older leaves turning purplish or dull blue-green (sometimes leaves become dull bronze with purple or brown spots).

The soil usually contains 1,000 to 4,000 pounds/acre of total phosphorus (except sands, which may have less than 200 pounds/acre), but the great majority is tied up in soil minerals and organic matter and is unavailable to plants. Available phosphorus typically runs from 30 to 90 pounds/acre. Most crops need 10 to 50 pounds/acre per year, so some soils can naturally supply a season's crop needs. Additional available phosphorus can be slowly released from mineral and organic matter reserves by microbial action (see Chapter 5). Soil phosphorus does not leach, but erosion of topsoil can pollute water with phosphates.

The best way to supply supplemental phosphorus (which is especially important during early growth when root systems are established) is by using a mined phosphorus source, which is a slower release material, along with a high quality fast-release source. Most dry commercial fertilizers only use the soluble form, diammonium phosphate (DAP), and liquids usually contain phosphoric acid or other available forms. The problem with totally soluble phosphorus fertilizers is that soon after they are in the soil (in a matter of days), the soluble (available to plants) phosphorus becomes tied up and insoluble (unavailable except very slowly by natural processes). Thus, most if not all of the phosphorus in a preplant starter fertilizer is "wasted" as far as that year's crop is concerned (how much can a small seedling plant grab?). Nearly all will go into the soil's phosphorus reserves.

In the soil, phosphorus that is "tied up" by combining with calcium is more easily released for plant use than is phosphorus combined with iron or aluminum. Therefore, a soil with high calcium levels is better than a low-calcium, acidic soil, and the natural-mined phosphorus sources that also contain calcium (rock phosphates) are good slow-release materials, especially in soils with high biological activity.

In fertilizers, the phosphorus content is usually given as "phosphoric acid" or "phosphate" (P_2O_5 equivalent), while scientific studies and soil tests may deal with phosphorus in an elemental basis. To convert from one into the other: $P = P_2O_5$ 0.437 or $P_2O_5 = P$ 2.29. Actually, the phosphorus in the fertilizer is not pure elemental phosphorus nor P_2O_5; it is usually dihydrogen phosphate (H_2PO_4; orthophosphate) or hydrogen phosphate (HPO_4).

Phosphorus Fertilizer Sources:

1. Rock phosphates. Natural mined deposits, from which other phosphorus fertilizers are made. They contain other useful elements, including trace elements. There are different kinds, which vary in effectiveness, including:

   a. Hard rock phosphate (24 to 30 percent P_2O_5, up to 30 percent calcium). Very low solubility; only useful as a long-term soil corrective.

   b. Idaho phosphate (27 percent P_2O_5, 25 - 28 percent calcium). Better solubility; a good slow-release source. Can be broadcast or used in starter blends. Also good as a manure additive (conserves nitrogen). With six to seven percent carbon.

   c. North Carolina rock phosphate ("reactive" phosphate) (30 to 33 percent P_2O_5, 33 to 35 percent calcium). Good solubility in acid soils; can give good crop response.

   d. Colloidal phosphate (soft rock phosphate) (18 to 22 percent P_2O_5, 16 to 19 percent calcium). A mixture of rock phosphate and clay. Lower solubility, so a 50 percent higher application rate may be needed. Best availability in acid soils.

2. Ordinary superphosphate (0-20-0) (16 to 22 percent P_2O_5, 18 to 21 percent calcium, 11 to 12 percent sulfur). A mixture of calcium phosphate and gypsum (calcium sulfate). Good solubility; makes a good side dressing. Little used today.

3. Triple superphosphate (0-46-0) (44 to 52 percent P_2O_5, 12 to 14 percent calcium). Good solubility, but does not give good plant response. Not recommended for biological farms.

4. Monoammonium phosphate (11-52-0, MAP) (48 to 53 percent P_2O_5, 11 percent nitrogen). High solubility; excellent crop response. A high quality soluble phosphorus source for starters (low pH, low ammonia); also supplies nitrogen.

5. Diammonium phosphate (18-46-0, DAP) (46 percent P_2O_5, 18 percent nitrogen). The most common soluble phosphorus source in dry blended commercial fertilizers. High solubility, but poor crop response (high pH, releases toxic ammonia, burns roots). Also supplies nitrogen. Not recommended for biological farms.

6. Liquid fertilizers (variable, made with phosphoric acid, orthophosphates and polyphosphates). Polyphosphates are a better source because they give slower release. High solubility; good as liquid starters. Some forms of phosphoric acid contain harmful impurities.

7. Basic slag (3.5 to 8 percent P_2O_5, 32 percent calcium, plus iron, zinc, manganese, copper and molybdenum). By-product of steel manufacture. Low solubility; high pH. May contain some heavy metals. Not a preferred source.

8. Bone meal (21 to 30 percent P_2O_5, 0.7 to 4 percent nitrogen). Slow availability. Used mainly by organic gardeners or vegetable farmers.

9. Animal manures (variable, 0.2 to 1.3 percent P_2O_5, plus nitrogen, potassium and other elements). Poultry manure is highest in phosphorus. The best practice for adding phosphorus to soil when you have animal manures is to mix rock phosphates with the manure before spreading. The rock phosphate ties up (saves) nitrogen, and the mineral form of phosphorus is also made more available by microbial action.

Potassium

Plant use of potassium is second only to nitrogen in amount, approximately 150 pounds/acre per year for most crops. Potassium is needed by plants for various metabolic activities, including enzyme functions, water use, balancing electrical charges in cells, and energy release. Crops with adequate potassium grow well, have strong stalks (resist lodging), produce a lot of sugar and protein, mature early and resist diseases. Potassium is needed throughout the growing season and is mobile within the plant. Symptoms of potassium deficiency include slow growth, weak stems and the edges of older leaves turning yellow, then brown, especially between the veins.

Most soils have large amounts of total potassium (30,000 to 35,000 pounds/acre), except for sandy soils, which may 2,000 pounds/acre or less. But, over 90 percent of the total potassium is tied up and unavailable to plants. About 100 to 150 pounds/acre are available at any one time, but only five to 15 pounds/acre of that is easily available (dissolved in the soil solution). The rest is held on and between the microscopic plates that make up clay particles, so it is sometimes called *slowly available*. A very small amount of the soil's unavailable potassium can be released by microbial action, but not enough to feed a crop for a year, according to textbooks (although in biologically active soil, the amount of readily available potassium does often markedly increase through the years; see Chapter 9, "Monitoring Soil Fertility Changes").

On fertilizer bags and soil tests, potassium is usually expressed as potassium oxide, K_2O (often called potash). The conversion between the oxide and elemental potassium or vice versa, is: $K = K_2O$ 0.83 and $K_2O = K$ 1.2. The potassium in fertilizers or organic matter is not potassium oxide, but other potassium compounds or potassium cations.

Over-fertilization with potassium may give high yields, but generally results in luxury consumption by plants, with the result that too much potassium is present in feeds for animals. Both in the soil and in animal nutrition, too high levels of potassium "crowd out" calcium and magnesium, resulting in deficiencies of these cations. In dairy cattle, low magnesium and calcium and high potassium can lead to serious health problems including "grass tetany" (hypomagnesemia) and milk fever (hypocalcemia). Once again, the balance of elements is very important in biological agriculture. The common recommendation to apply a couple hundred pounds/acre or more of potassium fertilizer to alfalfa fields is short sighted and does not consider the end result of unbalanced fertilization.

Potassium Fertilizer Sources:

1. Potassium sulfate (0-0-50-17S) (50 percent K_2O, 17 percent sulfur). With two needed plant nutrients; does not harm roots or soil life. Good plant availability, meaning lower rates can be used. Works well as a side dress in high-magnesium soils. The preferred form of quick-release potassium for most soils. It is lower in solubility than potassium chloride.

2. Potassium chloride (0-0-60 [muriate of potash — red color] or 0-0-62 [Kalium potash — white color]) (60 to 62 percent K_2O, with the rest being chloride). These are strong salts, not suitable in starters, since they may burn roots or kill soil life. Should not be used on chloride-sensitive crops, including potato, tomato, tobacco, corn, soybeans, cucumber, pepper, blueberry, strawberry and avocado. The high chloride content

supplies more chlorine than plants need. The excess must leach out of the root zone, but while leaching it can take needed calcium with it. A build-up of chloride can occur in poorly drained heavy soils. There have been reports lately of reduced yields and poor stands of soybeans from over use of potassium chloride. Should *not* be used in biological agriculture. If for cost reasons it must be used, do not exceed 100 pounds of 0-0-60 per acre per year, and bulk spread it in the fall.

3. Potassium magnesium sulfate (0-0-22, sul-po-mag, sulfate of potash-magnesia, K-mag) (22 percent K_2O). Also with 22 percent sulfur and 11 percent magnesium. Supplies three plant nutrients. Good availability; a good source where soils are low in magnesium.

4. Potassium nitrate (13-0-44) (44 percent K_2O, 13 percent nitrogen). Provides two plant nutrients, but is seldom used because of high cost.

5. Greensand (variable, 6 to 8.4 percent K_2O). A natural mined product, also called glauconite. Also contains some magnesium, iron and other trace elements. Slow release; often used by organic farmers and gardeners.

6. Granite dust (variable, 3.6 to 6 percent K_2O). A natural mined product; also with some magnesium, iron and other trace elements. Slow availability; a good slow-release source often used by organic farmers and gardeners.

7. Kiln dust (variable, 0.5 to 5 percent K_2O). A by-product of cement and burnt lime manufacture; also contains 28 to 36 percent calcium, 2 to 6 percent sulfur and trace elements. Fairly fast action; good plant response.

8. Animal manures (variable, 0.3 to 2 percent K_2O). Good slow-release source of several plant nutrients if applied in moderate amounts or composted (fields closest to the barn are often too high in potassium). Also adds organic matter and improves soil structure.

9. Kelp (seaweeds) (variable, 4.8 to 15.6 percent K_2O). Excellent natural source of many elements (especially trace elements) and plant growth-promoting substances.

10. Wood ashes (variable, 7 to 8.6 percent K_2O). Also supplies a little phosphorus and trace elements. Good availability; alkaline. Mainly used by gardeners and organic growers.

11. Sawdust (variable, 0.004 to 1.4 percent K_2O). Also supplies a little nitrogen and trace elements. Slow release; mainly used by gardeners and organic growers.

Calcium

Although it is called a secondary element, some legume crops take up about 100 to 250 pounds/acre per year of calcium. Corn and grass crops use about 15 to 40, and fruits and vegetables from 15 to 200 pounds/acre. Calcium has many important functions in the plant. It strengthens cell walls and is necessary for tip and root growth. It plays a key role in cellular membrane functions and the transfer of materials in or out of cells. It controls several metabolic enzymes and some processes in cellular respiration (energy release). It is needed for normal nitrogen use and protein production. It improves fruit quality and increases disease resistance. Calcium is not readily transported from one part of a plant to another, so it is very important that a growing crop have a readily available supply throughout the growing season. Calcium-deficient plants show distorted, curled and/or yellow to brown young leaves, poor root growth, premature flower drop, poor seed set and poor quality fruit (susceptible to rot diseases).

Most soils have moderate to high levels of total calcium. Sandy and silt loams may have about 4,000 to 5,000 pounds/acre, while clay loams have around 18,500 pounds/acre. Much is tied up in soil minerals such as calcitic limestone, dolomite, gypsum, apatite, calcium feldspars and amphiboles. Many soils in the western United States have high calcium levels, but in the eastern United States where rainfall is higher, soils tend to be leached of much calcium, and crop deficiencies can occur. Weathering, roots and microorganism activity will release some of the tied-up soil calcium (about 20 to 60 percent of the total can be exchangeable or in solution). Since other cations (magnesium, potassium, sodium) interfere with calcium absorption when they are too high, best crop uptake occurs when the percent base saturation of calcium is high (70 to 85 percent). In fact, when soils are in balance (see Chapter 6), plants need less calcium than in out-of-balance soils. Often soils that show adequate calcium on a standard soil test still do not have enough readily available calcium to provide a crop's needs all season long. In these cases, supplementation with extra calcium will pay off, since it is not an expensive element to buy. In high pH soils, at least some of the supplementary calcium should come from gypsum (calcium sulfate) or a BioCal type of product with high soluble calcium.

Other benefits of relatively high soil calcium include improved soil structure (it flocculates, or clumps colloidal particles) and stimulation of beneficial soil organisms, including nitrogen-fixing bacteria and earthworms. With good biological activity, organic matter decay is increased, nutrient release improved, and humus levels increased. Over all, crop growth and quality usually improve. Plant uptake of other elements also increases, giving more nutritious food.

Calcium-Containing Materials:

1. High-calcium lime (calcium carbonate, calcitic limestone, calcite, aragonite) (variable, 30 to 40 percent calcium). Less than 3 percent magnesium, plus small amounts of trace elements. A liming material (raises pH; see Chapter 10). Fairly low solubility; best worked into upper soil under acid soil conditions.

2. Dolomitic lime (calcium magnesium carbonate, dolomite) (variable, about 22 percent calcium, 8 to 20 percent magnesium, plus small amounts of trace elements). A liming material (raises pH). Low solubility; best worked into upper soil under acid soil conditions. Not always an effective plant source of calcium. Not a good source for high-magnesium soils.

3. Marl (calcium carbonate + clay impurities) (variable percent of calcium; an impure form of high-calcium lime). A good source in areas where it is available; would require higher application rates than pure lime. Low magnesium.

4. Kiln dust (variable, 28 to 36 percent calcium, plus 0.5 to 5 percent potassium, 2 to 4 percent sulfur and some trace elements). A by-product of cement and burnt lime manufacture. Very fine particles, can be surface applied. Good plant response. Low magnesium. The highest soluble calcium source that we have found on the market.

5. Other industrial by-products, including paper mill sludge, sugarbeet waste, water softener by-product and oyster shell. These contain mostly calcium carbonate, plus variable amounts of impurities. Check for possible toxic heavy metals or other problem substances.

6. Liquid lime (a suspension of fine particles of any liming material) (variable calcium content; can have high magnesium). Good availability, but expensive because of water content. Can be surface-applied.

7. Gypsum (calcium sulfate) (22 to 23 percent calcium, 18 percent sulfur). Moderately low solubility; best worked into upper soil. Supplies both calcium and sulfur. Not a liming material (does not raise pH); therefore good for adding calcium to high pH soils. Should not be used on low pH soils or if the percent base saturation of calcium is below 60 percent.

8. Rock phosphates (see under "phosphorus" for details) (variable, 16 to 30 percent calcium, plus phosphorus and trace elements). Low solubility; best worked into upper soil. Not liming materials (do not raise pH).

9. Ordinary superphosphate (0-20-0) (18 to 21 percent calcium, plus 16 to 22 percent phosphate, 11 to 12 percent sulfur and small amounts of trace elements). Moderate solubility; can be surface applied. Not a liming material (does not raise pH).

10. Triple superphosphate (0-46-0, concentrated superphosphate) (12 to 14 percent calcium, 45 to 47 percent phosphate). Good solubility, but not always a good plant response. Not a liming material; often makes soil very acid in time.

11. Calcium nitrate (15-0-0) (20 to 21 percent calcium, plus 15 percent nitrogen). High solubility; can be used as a foliar. Expensive; mainly used on high-value crops as a high quality nitrogen source that also contains calcium. Used in the right conditions and with the right timing, it can give remarkable results.

Magnesium

Plants use moderate amounts of magnesium, typically from 15 to 65 pounds/acre in a season. In the plant, magnesium is part of the chlorophyll molecule, so is necessary for photosynthesis and sugar production. It also has enzyme functions and controls cellular respiration (energy release), starch translocation, and protein and oil production. Unlike calcium, magnesium can usually be transferred from one part of the plant to another. Plants deficient in magnesium develop a pale yellow or white color on older leaves between the veins; later these leaves may turn brown and die from the edges inward. A few species of plants develop purplish older leaves at first. Low magnesium in crops can lead to animal health problems.

Most soils contain from 6,000 (silty loams) to 10,000 (clay loams) pounds/acre of total magnesium (sandy soils may have 2,500 pounds/acre or less). Soils that develop from magnesium-rich rocks can have very high amounts. Most soil magnesium is tied up in minerals, but natural release can make from 100 to 800 or more pounds/acre available to plants per year. Because too high calcium, potassium or sodium can reduce plant uptake of magnesium, some soils may not supply adequate amounts to a crop. Sandy soils, with a low cation exchange capacity, can easily be deficient. For most soils, a percent base exchange of 12 to 18 percent for magnesium is good. High magnesium levels are undesirable not only because they exclude calcium, but also because in some soils higher magnesium tends to "cement" clay particles together, leading to crusted or "tight" soil. Both magnesium and sodium ions tend to "collapse" soil structure, while calcium improves it. Therefore, I like to see 70 to 85 percent base saturation of calcium, and a calcium to magnesium ratio of 5:1 to 7:1. See Chapter 6 for more on soil balance and Chapter 10 for calcium:magnesium ratios.

Magnesium-Containing Materials:

1. Dolomitic lime (calcium magnesium carbonate, dolomite) (variable, 8 to 20 percent magnesium, plus about 22 percent calcium and some trace elements). Low solubility; not always an effective magnesium source (may be cheap in some areas, but little nutrient release for 18 months or more). A liming material (raises pH).

2. Sul-po-mag (0-0-22-23S, potassium magnesium sulfate, sulfate of potash-magnesia, K-mag) (11 percent magnesium, plus 22 percent K_2O and 23 percent sulfur). Good solubility; a good fast-release source of magnesium, potassium and sulfur.

3. Magnesium sulfate (Epsom salts) (9.6 to 10.5 percent magnesium plus 11 to 14 percent sulfur). Very soluble; a good fast-acting material in foliars.

4. Magnesium oxide (60 percent magnesium). High solubility with lower cost. It is not a liming material, so it does not affect pH.

5. Basic slag (open hearth) (3.4 percent magnesium, plus 3.5 to 8 percent phosphate, 32 percent calcium and several trace elements). Slow-release; alkaline.

Sulfur

Sulfur is needed to make quality, complete protein (it is part of three amino acids, cystine, cysteine and methionine). Sulfur is part of an enzyme used in metabolizing nitrates, so without adequate sulfur, excess nitrogen will build up as nitrates in plants, with potential animal health problems. Sulfur also is part of the B vitamins thiamine and biotin, and of coenzyme A, which is involved in cellular respiration and fatty acid metabolism (sulfur increases oil production in crops). It is needed for chlorophyll formation, root growth and nitrogen-fixing root nodule bacteria. Sulfur is sometimes called "the neglected element."

Sulfur is called a secondary element, but most crops take up about as much sulfur as they do phosphorus. Eight tons of alfalfa takes up 46 pounds, 150 bushels of corn contains 10 pounds (or 25 pounds in 15 tons of corn silage) and 60 bushels of soybeans has 12 pounds of sulfur. Plants mainly absorb sulfur as the sulfate ion (SO_4^-).

Sulfur in the soil can come from either sulfate that weathers out of minerals or from decayed organic matter. In good biologically active soil, from 70 to 90 percent of the soil's total sulfur is in organic matter. About 2.8 pounds/acre of sulfur are released each year for every 1 percent of soil organic matter. About 55 percent of the sulfur in raw manure is released the first year. Sulfate ions readily leach from soil, although in high-magnesium soils, it can take some of the excess magnesium with it. In water-logged soils, sulfate or organic sulfur can be converted into toxic sulfides by bacteria. Iron sulfide will remain in the soil, while hydrogen sulfide is a gas that can escape into the air.

Sulfur used to be found in many older fertilizers based on mined rocks, and some came from air pollution. But with the use of pure N-P-K fertilizers and lower sulfur air pollution, many crops are low or deficient in sulfur. Harvest removal and little recycling of organic matter can lead to low soil sulfur.

The ideal carbon:nitrogen:sulfur ratio in the soil is about 100:8:1, or as much as a 10:1 nitrogen:sulfur ratio. Soil organic matter is in a delicate balance that can be changed by many factors. The balance between nitrogen and sulfur in the soil system is critical as to how much humus is retained in a soil. Soil microbes will keep decomposing organic matter (crop residues, manures) until the N:S ratio is approximately six or 7:1. Another ratio of importance is the C:N ratio, which is approximately 10 or 11:1 in stable organic matter (humus), since humus is about 60 percent carbon.

Translating these ratios into something meaningful indicates that for every pound of sulfur available for soil microbes to use, the soil will be able to retain a little over 100 pounds of humus. Without that pound of sulfur, the 100 pounds of humus will be decomposed by microbes and lost into the air as gases. Here is how it works:

One pound of S complexes with six to seven pounds of N.

Six to seven pounds of N complex with 60 to 70 pounds of C.

60 to 70 pounds of C complex with 100 to 125 pounds of humus.

Crop residues contain 0.05 to 0.15 percent sulfur, or one to three pounds of sulfur per ton, only enough to save five to 15 percent of the residue as humus.

The best way to ensure that you maximize soil organic matter is to apply a sulfur source such as BioCal, gypsum or 100 pounds of ammonium sulfate for every ton of crop residue produced. Ammonium sulfate should be broadcast on top of the residue, and both should be tilled in lightly for maximum humus development. Use a sulfate material for best plant availability. Elemental sulfur can be too acidic for some conditions and must go through biological conversion to become available.

Research has shown that if the carbon-to-sulfur ratio in organic residues is above 50:1 (high carbon, low sulfur), most of the sulfur they contain will be immobilized in microbial cells, while below a 50:1 C:S ratio (lower C, higher S), the sulfur will be mineralized (transformed into plant-available sulfate; however, sulfate sulfur is leachable). The exact critical value of the C:S ratio does vary with different organic materials.

Sulfur-Containing Materials:

1. Elemental sulfur (0-0-0-100S) (100 percent sulfur). Not usable by plants unless converted to sulfate by soil bacteria, which may take at least two to four weeks for finely ground sulfur (therefore it should be worked into upper soil). Insoluble in water. Acidifies the soil, so useful for alkaline soils that need sulfur.

2. Ammonium sulfate (21-0-0-24S) (24 percent sulfur, plus 21 percent nitrogen). Fertilizer grade materials are industrial by-products; some may be harder to handle than others. High solubility. Large applications can acidify soil.

3. Ammonium thiosulfate (12-0-0-26S) (26 percent sulfur, 12 percent nitrogen). A 60 percent water solution which can be directly soil-applied or used in irrigation systems. It is compatible with liquid nitrogen and most N-P-K solutions.

4. Sul-po-mag (0-0-22-23S, potassium magnesium sulfate, sulfate of potash-magnesia, K-mag) (23 percent sulfur, 22 percent K_2O and 11 percent magnesium). A good source if soil needs magnesium.

5. Magnesium sulfate (0-0-0-14S) (14 percent sulfur, 10 percent magnesium). Very soluble; good in foliar sprays if magnesium is needed also.

6. Potassium sulfate (0-0-50-18S) (18 percent sulfur, 50 percent K_2O). High quality source of potassium plus sulfur. See above under "potassium" for more detail on potassium sources.

7. Ordinary superphosphate (0-20-0-14S) (14 percent sulfur, 20 percent phosphate, 18 to 21 percent calcium, plus some trace elements). A good material, but not used much today.

8. Calcium sulfate (0-0-0-17S, gypsum) (17 percent calcium, 22 to 23 percent calcium). Fairly low solubility; best worked into upper soil. A good material at moderate rates for high-pH soils.

9. Kiln dust (variable, 28 to 36 percent calcium, plus 0.5 to 5 percent potassium, 2 to 4 percent sulfur and some trace elements). A by-product of cement and burnt lime manufacture. Very fine particles, can be surface applied. Good plant response. Low magnesium.

10. Basic slag (variable, can contain 0.2 to 3 percent sulfur, depending on type). A by-product of steel manufacture. Also supplies several other elements, especially phosphorus. Alkaline.

11. Animal manures (variable, 0.1 to 0.2 percent sulfur). Also supply many other elements, plus organic matter. Use in moderate amounts; work into upper soil.

12. Animal by-products, such as tankage and bone meal (variable, 0.2 to 0.4 percent sulfur). Also supply other elements, especially nitrogen.

13. Sewage sludge (can contain up to 0.4 percent sulfur). Also supplies many other elements. Check for possible toxic heavy metals or other substances.

14. Plant residues and by-products (variable, 0.1 to 0.5 percent sulfur; higher in high-protein crops such as legumes). Sources include crop residues, green manures, cannery wastes, soybean meal and compost. Supply many other elements, plus organic matter. Apply at moderate rates; work into upper soil.

Trace Elements

The trace elements (also called micronutrients) considered essential for all plants include iron, manganese, zinc, copper, boron, molybdenum, nickel and chlorine. The latter three are not normally supplemented or tested for, since many soils have plenty.

Each trace element has different specific functions in plants, but rather than bore you with the details, let's just say that in general, the trace elements function in various plant metabolic activities, usually as enzyme activators. Without them, the plant's cellular "machinery" would shut down. Often, an adequate supply of a certain trace element can make the difference between a high quality crop and a so-so one.

Many soils contain some of the trace elements at levels sufficient for many years' worth of crops, but with the use of pure N-P-K fertilizers without trace elements, and if organic matter is not recycled, the available soil supply can become deficient. A soil test should reveal any glaring problems. In a biological farming system, the best approach is to first get the major and secondary elements into balance; use natural-base fertilizers (ones that include mined rock materials); and feed the crop a balance of all elements, including trace elements.

A relatively high calcium level, good soil aeration and high humus should improve availability and balance of trace elements. Humic and organic acids from humus and microbes act as chelating agents causing trace elements to be held in forms that are more readily available to plants than they would be otherwise. Row-feeding, top dressing and foliar feeding are the most efficient ways to supply trace elements without completely changing soil levels. The sulfate forms of trace elements work well in dry starter fertilizers and for broadcast corrective treatment. The more expensive chelated forms are essential for liquid and foliar application. Oxide forms of trace elements have very low solubility, and I do not recommend them.

Trace Element Fertilizer Sources:

1. Zinc. Essential for corn starters.
   a. Zinc sulfate (36 percent zinc). You generally need about a five pound/acre application to supply the crop and build soil levels.
   b. Zinc chelate (8 to 10 percent zinc). Apply one to two quarts/acre in liquid starters.

2. Manganese. Another element essential for corn starters.
   a. Manganese sulfate (28 percent manganese). Apply 10 to 15 pounds/acre during soil corrective periods.
   b. Manganese chelate (5 percent manganese). Apply one to two quarts/acre as liquid.

3. Iron. Deficiencies occur on high pH soils (above 7.0) and on high-phosphorus soils.
   a. Iron sulfate (20 percent iron).
   b. Iron chelate (5 to 15 percent iron). Good for foliar feeding on high pH soils.

4. Copper. Deficiencies can be corrected with soil applications or chelated foliar sprays. Copper is contained in some fungicides, and adequate copper in the soil may help prevent some fungal diseases.
   a. Copper sulfate (25 percent copper). Use four to 10 pounds/acre per year as a bulk spread.
   b. Copper chelate (7 to 13 percent copper). Use one to two quarts/acre as liquid.

5. Boron. It has received more attention than other trace elements. Alfalfa requires a lot and, in some areas, soils are deficient. Boron leaches, so some should be added each year. Too much boron can be toxic, therefore broadcast application is the safest. One to two pounds of boron (figured on an elemental basis) per year is a maximum application. There are many dry and liquid forms, including:
   a. Borax (11 percent boron).
   b. Boron frits (10 to 17 percent boron).
   c. Solubor (20 percent boron).
   d. Calcium borate (10 percent boron).

13

TAKING ACTION

You have information on soil testing, reading a soil report and on fertilizers. This is the chemical (nutrients) part of soils, an area where more advice is given because it is easier to measure progress. The physical and biological parts of the soil are not easily measured. There are not as many places where you can send soil samples to get advice about these two areas, or even an evaluation. In this chapter, I will give suggestions for all three areas. Common sense, good evaluation, understanding how soil works, and practical suggestions are my way of teaching better soil stewardship.

It is impossible to list all the examples of soil types, soil correctives, fertilizer programs and methods of working with the physical and biological properties of a soil in a manual. After reading and studying this approach, if you still have questions or you are not sure, contact Midwestern Bio-Ag for some consulting help at 1-800-327-6012.

The first two examples I will use are from my own farm (Otter Creek Organic Farm, Lone Rock, Wisconsin) and the Bio-Ag Learning Center across the road. I am very familiar with these farms, and I can tell it as I see it from personal experience. The other examples are from Midwestern Bio-Ag consultants from all over the country, not only to show different soil types, but that the approach will be different in different areas for different crops and soils.

A program can be put together for any farmer anywhere. Look at where the farm currently is and what the farmer is doing now. Identify limiting factors and take action. I have yet to come across a farm which doesn't have limiting factors and need improvements in some areas.

How Do I Start?

In most cases success does not come from buying a pail or bag of something, but from changing some management practices, along with a good soil fertility program. Below I have a list of what biological farmers consider negative practices. Start by eliminating or reducing the negatives and adding more positive practices.

Negatives:
- salt fertilizers
- too much nitrogen use
- excess manure use, or excess fertilizer use
- use of any material that causes a soil to get out of balance
- an imbalanced fertilizer
- too much highly soluble fertilizer
- anhydrous ammonia
- herbicide overuse
- insecticides
- too much tillage, over-aeration
- tillage at the wrong time
- never using green manure crops
- burying residues below the aerobic zone
- working soils too wet
- no tillage when a soil needs air and drainage

I am sure the list can go on and on. Evaluate your soil. If everything is ideal with lots of organic matter, lots of soil life, beautiful tilth and structure, abundant supply and balance of all minerals, and healthy, high-yielding crops, maybe you have already eliminated the negatives.

Now let's look at adding positive practices and doing things to make soil better. Following is a list of those practices. Some may or may not be able to be accomplished with the soils in your area.

Positives:
- evaluate your soil test and remineralize deficiencies with correct materials for your soil
- use a balanced crop fertilizer containing all the essential nutrients that allows maximum crop performance and root growth
- incorporate residues shallowly (chopped into small pieces) for rapid decay in the aerobic zone
- use deep tillage to ensure proper drainage
- create an ideal home for soil life by keeping their food on top and by not leaving the soil surface bare and exposed
- grow green manure crops whenever possible

- use livestock manure (if available) to correct soil shortages; if all fields have similar levels of minerals, use light coats of manure on many acres
- use compost or humates if available

As with the negative list, many more things can be added to this positive list. Biological stimulants, kelp, enzymes, humic acid and bacterial inoculants can be useful, providing all the other beneficial practices (see Chapter 2) are followed as well.

The Soil Test — Evaluation and Action

The best way for me to interpret the information on a good soil test is by using an actual case study. I will give a farm history, show the soil test, outline the soil corrective program, outline the crop fertilizer program, look at the crops grown and results obtained, follow up with a soil test after three years, and outline the present program.

Case Studies

Farm History — Otter Creek Organic Farm

A 240-acre farm located in southwestern Wisconsin, with 120 acres of cropland and 100 acres of good creek-bottom pasture. It was a dairy farm until the dairy buy-out in the early 1980s. Until then crops were rotated, manure was applied, no commercial fertilizers were used, and only a minimum amount of herbicides and no insecticides were used. Lime had not been used for 35 years. The tillage was plowing, disking, digging and cultivation with a digger-type cultivator.

After selling the cows in the early 1980s, corn and soybeans were grown in rotation. A small amount of liquid fertilizer was used as a starter, and a dry blend was bulk spread at 150 pounds per acre.

In 1994, the farm ownership was transferred, and it became Otter Creek Organic Farm. Soil tests were taken, and a complete evaluation was done. Foxtail and velvetleaf were the main weeds. It was a wet year, and gullies from erosion had to be filled in. Herbicides had been used on the soybeans, and weed control was good. Erosion was a problem. Earthworms were rare, and soil was crusted with poor residue decay.

The land is gently rolling to level. The soil type is mostly silt loams, with some sandy knolls and some areas with clay knobs. The 120 acres of cropland are now divided into four fields. For this example, I will just take one field, look at the soil test, put together the recommendations and follow up with the future test.

The Action

The decision was made to farm using organic methods, which meant no herbicides, insecticides or commercial fertilizers. Normally, if you want to farm organically, I recommend that you first get the crop rotation set up, balance the minerals and have a biologically active soil. It is very difficult to go from low-testing, "dead" soils which are low in organic matter to successful organic farming. I believe you should get the soil healthy, balanced and mineralized first. Then you have earned the right to farm it organically and you should be successful.

In this case, as I have already pointed out, the previous farmer had used no insecticides, no anhydrous ammonia, and only minimal herbicides; however, to build soil fertility, improve soil structure and control weeds, inputs would be required, plus a small grain or forage crop in rotation seemed essential. Dividing the farm into four fields was the plan. The rotation was to be corn, soybeans, seeding a small grain as a nurse crop with alfalfa, clover and perennial rye grass. The last crop in the rotation would be a first cutting of alfalfa, and then back to soybeans. Due to a limited budget, a major fertility correction would be made on one-fourth of the farm each year. The land being seeded down to forages would get this correction, and at the end of four years the whole farm would have one round of correction. Soil tests would then be retaken and additional correction applied as needed. Due to low fertility, row crops were to have applications of balanced row-applied fertilizer at higher than normal rates, at least until the major correction was made.

Following soybeans, rye grain would be fall-seeded as a green manure crop. The corn crop would be interseeded with clover, hairy vetch and annual rye grass. Anytime I could grow a crop for soil building, feed the soil life and bind up soluble soil nutrients from loss to erosion or leaching, I would. A major focus was that I did not want to see soils bare or miss an opportunity to capture the sun's energy by growing protective crops.

Soil tests done on this farm indicated a low fertility level. The calcium levels were too low and magnesium too high. Following is a table showing the fertility levels when we started, including tests two and four years later. Midwest Labs in Omaha did the testing.

Field 1, Otter Creek Organic Farm

Test	Fall 1993	Fall 1995	Spring 1997
CEC	8.9	8.8	10.7
OM	1.9 L	1.9 L	2.3 M
P1 ppm	22 M	45 VH	72 VH
P2 ppm	37 M	84 VH	156 VH

Test	Fall 1993	Fall 1995	Spring 1997
% Base Saturation			
K	2.2%	2.9%	2.6%
Mg	32.6%	27.9%	27.7%
Ca	65.1%	69.2%	69.6%
ppm			
K	78 L	99 M	108 M
Mg	355 VH	295 VH	372 VH
Ca	1,180 M	1,225 H	1,540 H
pH	6.9	7.2	7.4
ppm			
S	9 L	8 L	13 L
Zn	1.7 M	2.4 M	3.8 H
Mn	25 H	16 H	14 M
Fe	32 VH	30 VH	26 H
Cu	0.5 L	1.1 M	1.5 H
B	0.7 L	0.8 M	1.2 M

Note: L = low, M = medium, H = high, VH = very high

Doing this testing and looking over the results does give me confidence in soil testing. We do make sure we get a good representative sample. The spring 1997 test had a combination of three samples. We were checking to see if different areas and soil types in the field tested differently. Our results showed very little fertility difference.

After the 1995 test, we found a source of chicken manure from a laying hen operation. Not wanting to fertilize weeds and battle them all summer, we chose to put on smaller amounts of chicken manure, preferably in the fall. We use either corn stalks to balance the excess nitrogen from the manure or a green growing crop to utilize it and store it for future use. This also cuts down on the soluble nitrogen, reduces weed feeding and cuts down on the erosion of nitrogen and leaching.

Now that the soil fertility is at its current level, we will only use composted manure, and only for high-nitrogen-using crops such as corn, or to increase soil fertility if the present high levels drop.

Back to 1993. Let's look at the soil test and some fertilizer options. In balancing a soil, I like to start with calcium and phosphorus. If the pH were low, I would add the natural rock phosphorus first, at 500 to 700 pounds per acre, wait a season and then add the calcium. A low soil pH would speed the breakdown of the phosphorus and still provide calcium for the crop. Idaho phosphate contains 28 percent P_2O_5 and 30 percent calcium. This is my preferred source. This practice would not increase the pH, consequently I would still need to lime. In this case, because the soil is high in magnesium, I would use a high-calcium lime. Remember, in balancing a soil, if the cation element is in the VH range (very high), don't add more. These soils don't have a low pH, and because I wanted to fix it fast, I added both high-calcium lime and BioCal for extra soluble calcium.

The soil-correcting mix I used was:
- one ton high calcium lime
- 1,500 pounds BioCal (a soluble calcium, sulfur and boron source)
- 700 pounds Idaho phosphate
- 200 pounds potassium sulfate
- five pounds zinc sulfate
- five pounds manganese sulfate
- two pounds copper sulfate
- five pounds calcium borate

This is just one approach. It was over a $200.00 per acre investment (a one-time expense). Note that the same results can be accomplished by spending $40.00 per acre over five years as finances allow.

The minimum correction I would feel comfortable doing would be 300 pounds of Idaho phosphate (my best choice for this area; my next choice would be North Carolina rock phosphate, if price and availability were favorable). The Idaho phosphate is granulated, making it easy to blend with the other soil correctives. For my minimums in addition to the Idaho phosphate, I would also put on 500 pounds of BioCal, trying to balance the soil's calcium and phosphorus. I would still be short of minerals, but I could do the rest with a row or crop blend.

This field was seeded down with alfalfa and rye grass, so my crop fertilizer would be a blend that fits my soil, like 0-2-10-13S plus traces, or a custom mix. In this case, 200 pounds of potassium sulfate, two pounds of zinc sulfate, two pounds of manganese sulfate, one pound of copper sulfate and three pounds of calcium borate would be the minimum. If the crop were corn or soybeans, I could take a similar blend, reduce the rate slightly, row-apply and grow a good crop.

I use natural-base blends because the materials aren't all soluble when I put them on. Their mineral availability is spread over time as the crop needs them. In addition, natural materials contain elements which we may not have identified as essential and are needed in small amounts. An example is Idaho phosphate, which is an excellent source of molybdenum and comes free with the purchased phosphorus.

In addition to natural-based materials, I add kelp, fish, humates and shale, which are excellent mineral, enzyme, hormone, energy and vitamin sources. These are extras I'm sure are essential, but science has yet to confirm.

Where do I get these numbers of pounds of materials I use? There is no magic in the numbers used. They come from years of soil consulting on many farms. If the soil test indicates these nutrients are needed, add them. How much depends on your expectations and budget. Putting on the minimum keeps costs down, but also slows down results. I am over 50 years old. I can't go that slowly if I want to reap the benefits. I'm convinced the faster the fix, the faster the pay-back.

Supplying the major nutrients as a soil correctives and crop feeding the traces does bring up the soil test levels. When you feed a crop a balanced diet, this balance goes throughout the plant. As the plant residues decay, those high plant mineral levels go to feed the soil life to start cycling the minerals through the system and eventually increase soil exchangeable levels. That's why this process takes time. It didn't get this way overnight, and it won't be fixed overnight either.

Now let's look at the follow-up soil test (1995). I waited two years to re-test. Normally, I would do the major correction, keep applying the balanced crop fertilizer, feed the soil life and change tillage so as not to put limits on the soil structure, decay cycle and soil life — then recheck in four years. At that time I would have made the major correction on all fields. If the farm's soil test came back in the beginning with low pHs, I would do a lime correction on the whole farm and not wait until a field's turn came. In this example, since pH was fine but fertility low, I did use a higher level of a balanced row crop fertilizer so as not to starve the growing crop.

The 1995 test results did show that my efforts and expense were moving things in the right direction. To our surprise, the phosphorus really moved — more than doubled. Research in the past indicated that the material we put on was not soluble, wouldn't become available and was a waste of money. This research may be true for a hard, dead soil, but green manure crops, earthworms and microbes can do wonders if given a chance.

Note other soil changes. The magnesium came down, calcium up, potassium up slightly — all the things I wanted. The soil pH is coming up. This is an organic farm, and without a lot of manure and/or nitrogen use, the pHs will probably never be less than 7.0. Remember, pH is a measure of fertility; if I have the soil saturated with cations, the pH won't be low. Having a pH over 7.0 on a dead soil can lead to problems. If earthworms are abundant and soil microbes active, it's not a concern. They do the job of making minerals available, and I don't need to depend on acid soils for that job. Earthworms have a tendency to neutralize soil pH. To get the soil acid, I would need to kill them off, and I think that's unwise.

Sulfur and boron are anions and do not attach to the soil. Some are held by soil organic matter, but they can and do leach. The sulfur leaching removes some magnesium, as I wanted.

As for the other trace elements, they are increasing, except for iron. It was too high. I have not added any and have cut down on soil oxidation by less tillage and by not using harsh fertilizer materials.

The crops have been getting row-applied balanced fertilizers. As I mentioned before, in the fall of 1995, we found a chicken manure source and have been applying a light coat since then. After soybeans we plant fall rye grain for a green manure crop, and the corn is interseeded with different seed blends, mainly hairy vetch, red clover and rye grass.

Now it's 1997. What's left to fix and where? I sure don't need any more phosphorus. I do like high levels of phosphorus, since it affects crop quality and energy transfer. I see no disadvantage, since phosphorus does not leach. Phosphorus is an anion, but unlike nitrogen, sulfur and boron it has a triple negative charge and is held tightly to the soil cations, mainly by calcium if it is present. If the soil has a low calcium availability, the phosphorus will attach to iron and aluminum, greatly reducing its crop availability. The only way phosphorus can get off my farm is by crop removal and erosion. I will not be polluting area streams and lakes. With phosphorus this high in the soil, I expect very few insect problems or crop diseases. The only problem that could occur is zinc deficiency. If you move phosphorus up, you need to also move zinc and make sure it is crop-fed to the plant. I still need more calcium and will keep on applying 500 to 1,000 lbs. BioCal at least two out of four years. I like putting the BioCal on when I plant soybeans since they can use the extra sulfur to make better quality protein and they are a calcium-loving legume.

As for corrective blends, I will be using potassium sulfate at 150 to 200 pounds per acre along with a trace mineral mix; this all in addition to a balanced crop fertilizer. In a few years I will no longer need to do soil correcting and my fertilizer bill will go down. Our objective is not to get a perfect-looking soil test. We just want live healthy soils that are mineralized in a good balance.

How are the crops doing? In 1996, we took off the first crop of hay, averaging six 1,800-pound haylage bales per acre and sold them for $50.00 each. Then on June 10th (due to bad weather, it got too late) we planted soybeans for the tofu market. The crop was weed-free and did great, averaging 45 bushels per acre and was marketed at $18.00 per bushel.

In 1997 the crop was field corn. Weed control was excellent. The crop was good but not great. I have a yield monitor on my combine, and yields ranged from 129 to 215 bushels with test weights from 58 to 60. This range is due to soil structure, not fertility. The last two springs have been wet early in the season. As an organic farmer, we make five to six trips across the field to control weeds and to interseed. We are sometimes out there because we have to do the job, but conditions are often less than ideal, which can cause compaction.

In the fall we subsoiled and, in the future, by getting more soil life, better Ca:Mg ratios, and deeper and larger root systems, we should reduce the compaction problem. I expect corn yields in the 150 bushel range.

More work needs to be done. To fix a soil, you have to start on top in the aerobic zone. Now I can start to deepen that aerobic zone, and my goal is to have 12 inches of loose, well-aerated soil with a balance of minerals in the "high" range and lots of soil life.

There is no law in the organic farming book that says I need to starve the crop. Crops grown on healthy soils with lots of minerals do yield well. My best showing will likely come in a bad year when crops are doing poorly all over. These are the years when biologically farmed soils really out-perform conventionally farmed soils. Pour a lot of commercial fertilizer on, hope for moisture in the right amounts and at the right times, get the growing days . . . and you get yield. So many crop failures are blamed on the weather, when in reality it is soil conditions that are failing.

You see figures stating that it can take up to 500 years to build an inch of topsoil. I disagree. It takes that time when undisturbed. Louis Bromfield in the 1930s at Malabar Farm in Ohio said he could build six inches of topsoil in just five years. In his book, *Malabar Farm*, he outlined his procedures. In those days they didn't have the problems of high commercial fertilizer and chemical use as today. Back then it was the plow, poor management and poor fertility practices. They just didn't know. Plowing up and down the hills, removing crops and not putting manure or nutrients back until yields were terrible made farmers go under.

Bromfield's methods were seeding with oats and clover, liming, and using natural phosphates, hydrated lime, some manufactured fertilizers and trace elements. The first year he just cut the oats and clover and left them to lie on the surface as a mulch. In the fall, he added manure, more correctives and shallowly incorporated the residues. The next spring it was crop time, and he started his rotation with small grains, hay and some corn. After a few years of shallow incorporation, he would plow or chisel the soil to mix in more subsoil (hard, dead soil) with his healthy, active soil. After five years he had six inches of loose, crumbly biologically active mineralized soil. His reports of yield increases were almost unbelievable.

Here we are fifty years later, and biological farmers that are successful are using similar practices. Many now have a better understanding, but Mother Nature hasn't changed. It worked back then, and it will work now. It seems sad that after all these years, we still have so few biological farmers who understand what it takes, and are doing it. Chemicals came along, were cheap and gave wonderful responses in crop yields. In the beginning of commercial fertilizer use, many farmers rotated crops, grew a variety of plant species and even had biological activity in their soil. They needed fertility. The choices for fertility were based on price, and having no way to measure the decline in soil health, we ended up where we are today. That's why it takes so much more fertilizer to grow the same crops. If biological activity is low, if you have hard, tight soils with poor structure, and if the minerals are out of balance as is the case on many farms, it will take more fertilizer to grow the unbalanced crop. Remember the relationship between chemical, physical and biological soil properties. One method is pushing the soil, fighting against Mother Nature; the other is working with nature and letting the system do the work.

The 1998 season was the biggest challenge of all. We live in an area in the state which recorded an all-time record rain fall. Ten inches more than normal. Due to all the rain, soils were never ideal for working. Remember any condition less than ideal at planting is unacceptable. Your success for the whole year is dependent upon plants getting off to a good start. The field was planted late May for soybeans, heavy rains followed, rotary hoeing was impossible. I used the flame cultivator to suppress weeds early, but due to heavy rains weed control was marginal and the crop stand was poor.

So we started over with a shallow rotavation and replanting of beans on June 6th. The organic soybeans planted the second time were great. Weed control was also good. The field had a foliar demonstration plot using seaweed, humic acid and fish fertilizers. No yield response was noted; however, yields ranged from 65-71 bushel/acre. Maybe other benefits were obtained with the foliar applications, but nothing we could measure. The crop fertilizer was row applied at 225 lbs/acre using a custom mix of:

> 50 lbs. - Idaphos
>
> 50 lbs. - K-Mag
>
> 50 lbs. - K_2SO_4 (Potassium Sulfate)
>
> 15 lbs. - $ZnSO_4$ (Zinc Sulfate)

20 lbs. - $MnSO_4$ (Manganese Sulfate)

8 lbs. - $CuSO_4$ (Copper Sulfate)

10 lbs. - Calcium Borate

15 lbs. - Bio-Root (a kelp, fish root stimulating blend)

In addition to this, we bulk spread 500 lbs of Bio-Cal; the soluble calcium and sulfur liming material.

This may seem to you like a lot of materials, but as you can see from the soil test these nutrients are needed. Organic soybeans were contracted for $20 a bushel that year, and, considering the yields that were obtained, it certainly was a profitable investment. As the soil gets closer to balance, much less and possibly a slightly different blend will be used.

The year of 1999 will be a soil building year, growing green manure crops, applying more soil correctives, compost and no tillage. A year to enhance soil microbiological activity and soil structure.

Farm History — Bio-Ag Learning Center

My next example is from the Bio-Ag Learning Center. It is a farm I have worked since 1992. It is not organic, but we use no herbicides or insecticides. In our operation, with rotations, healthy soils and the crops we grow, insecticides are totally unnecessary. I don't believe that herbicides at low rates, used properly, are that bad. It's just that enough people are doing studies with chemicals, and we need more guidance on non-chemical farming. We really don't know if herbicides are good or bad, or which ones are safer. They are designed to kill, and I'm sure that weeds aren't the only species killed. Not using herbicides is certainly safe, and I know I'm not putting any yield limits on my crops from them. We have tools and knowledge today to do the job without chemicals. It's not easier. It takes more understanding and better management, but it's achievable.

The Bio-Ag Learning Center has 80 tillable acres and is divided into 37 fields. When we started farming it in 1992, I took soil samples from every field. I needed this starting point so I could make fertilizer decisions. This farm is across the road from Otter Creek Organic Farm, but was farmed totally differently. Both farmers plowed and both had dairy herds, but the farmer on the Learning Center land used more commercial fertilizer, as you will see from the tests. Both farms (as with many in Wisconsin) had high magnesium and low calcium soils.

The following table shows the results of soil tests taken in 1991, compared to samples from spring 1997. I chose two fields on the same part of the farm but one has been getting soil correctives and one has not. On the

Learning Center farm, we leave some fields untreated for future research. Both fields have been getting balanced crop fertilizer, so the only variation is the correctives. Again, testing was done by Midwest Labs in Omaha.

Bio-Ag Learning Center

Test	Ideal*	Field 4B 1991	Field 4B 1997	Field 8B 1991	Field 8B 1997
CEC	–	12.6	10.9	12.4	12.0
OM	3.0+	2.1	1.9	2.0	2.1
P1 ppm	50	23 H	23 H	32 H	60 VH
P2 ppm	100	34 M	36 M	39 M	124 VH
% base saturation					
K	3.2	2.1%	2.1%	1.8%	3.5%
Mg	below 20	33.8%	36.1%	32.5%	29.3%
Ca	75	64.2%	61.8%	65.7%	67.2%
ppm					
K	125	102 M	91 L	88 L	163 H
Mg	300	511 VH	471 VH	485 VH	421 VH
Ca	1,800	1,619 M	1,345 M	1,633 M	1,606 H
pH	–	7.2	6.9	7.2	7.3
ppm					
S	25+	26 VH	12 M	17 M	18 M
Zn	5	1.3 M	2.0 M	1.1 M	2.9 H
Mn	20	30 VH	26 H	22 H	12 M
Fe	20	24 H	34 VH	19 H	24 H
Cu	2	0.7 L	1.2 M	0.5 L	1.2 M
B	2	1.2 M	1.2 M	0.9 M	1.0 M

* Ideal = ideal balance level for these fields
Note: L = low, M = medium, H = high, VH = very high

On field 4B without soil correctives, we keep getting a crop. We run a tight rotation and do apply some livestock manure. The program for this field isn't taking me backward, but it certainly is not performing as field 8B does. Starting this year, field 4B will be put on a fertilizer corrective program and be seeded for a rotational grazing project. Note that on field 4B, just using a crop fertilizer with balanced minerals did improve trace elements zinc and copper. I'm not sure why the CEC is so much lower or the organic matter. Maybe we are using up humus or maybe it is just testing variations. It is a trend I certainly don't like.

Field 8B is used for our crop variety plots, so it is in a corn/soybean rotation. Yields generally run from 60 to 75 bushels per acre of soybeans and 150 to 210 bushels per acre of corn. Besides the row fertilizers, up to 50 units of extra nitrogen have been side dressed at cultivation, putting the nitrogen where and when it is needed. After soybeans, rye is planted, and the corn is interseeded with the grass/clover mix.

Two years ago we added six tons of compost per acre, but that was the only manure-type material used. Soil correction over the past six years has been 1,000 pounds of BioCal, one ton of gypsum, 500 pounds of Idaho phosphate, 12 tons of granite dust and three tons of high-calcium lime. The changes in this soil are noticeable. I'm not sure I need to add as much as I did, but I wanted to do the correcting with rock dust products. Those levels seem high, but that's over a period of six years — while growing wonderful crops.

The crop fertilizer for corn is 15-8-2-16S, with 3 percent calcium, 0.12 percent boron, 0.12 percent copper, 0.4 percent manganese and 0.5 percent zinc. This was applied at 350 pounds per acre. The difference between this fertilizer and the organic blend is that in this one we use ammonium sulfate, monoammonium phosphate and a manufactured homogenized trace mineral fertilizer. The trace mineral mix contains a base of high-sugar ammonium sulfate and North Carolina rock phosphate, with the trace elements acidified to a 5.5 pH using phosphoric and sulfuric acids.

I could have made these soil corrections on field 8B using potassium sulfate, monoammonium phosphate, Idaho phosphate and the needed calcium. In all my corrections, I like to use natural mined materials as part of the formula. I hope it is clear I am talking about two different but related topics: soil correction and crop fertilizers. This "in balance" is what a lot of this book's topics are about. Remember, a balanced soil (balanced chemically, physically and biologically) performs very well. Now add a balanced crop fertilizer designed for the specific crop needs and applied at specific times. Until soils are balanced, these crop fertilizers need to be altered to make up for soil nutrient shortages. A crop fertilizer should be adding an additional nutrient level above the soil's capacity to release minerals, providing a good balance for the soil "soup."

Soil "soup" is the solution of nutrients that are exchangeable or usable to the plant. The plant is a consumer of many minerals. It can consume them in a wide range of ratios. Alter your "soup" and you alter the uptake by the plant, its development and health, how the fibers are formed, cell walls, the plant's vitamin level, hormones and its own defense mechanisms. You can stimulate a pile, a volume of material, by having the soup be mostly N-P-K. Fertilizer efficiency is related to source of material and time of application.

Apply the soluble fertilizers early. That's how all fertilizers are marketed. How do you expect to have exchangeable nutrients left come mid-season

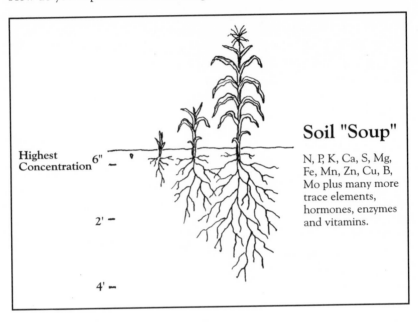

Soil "Soup"

N, P, K, Ca, S, Mg, Fe, Mn, Zn, Cu, B, Mo plus many more trace elements, hormones, enzymes and vitamins.

Highest Concentration 6" —

2' —

4' —

or when the plant really needs them? That's why the soil's microbiology and organic matter cycle are so important. The residues from the previous year should all be broken down by midsummer, thus releasing the nutrients contained in them. The bacteria eat first, then release.

Oh, you say, this is too complex. I will just give it a bath, fill up the soup pot with extras so I have plenty for the season. This is a common present-day approach to fertilizers with N-P-K: buy highly soluble stuff and put on enough. The problem is the side-effects. Soil microbiology is affected, along with mineral balance, root growth and plant health. There is a competition going on in the soil. Plants aren't good at selecting. A high level of one nutrient certainly affects the utilization of other minerals. The zinc-phosphorus relationship is a good example. Excess can cause as many problems as shortages. Farms, especially with a lot of livestock and manure, do have these problems. We need to add more of an element because the levels of some of the others are too high. Potassium and magnesium have this competitive relationship. If magnesium is high it takes more potassium to grow the same crop. If potassium is high it interferes with the plant's magnesium use. For example, grass tetany is due to high plant potassium and low magnesium. The list could go on and on — just another reason for soil balance.

With excess N, P, and K, soil leaching will certainly be heavier. Some of what you apply can attach to soil colloids or become biologically tied up in decay cycles. This excess N-P-K soup will also start to move down in the soil along with water. Nitrogen is the fastest to move, and because phosphorus has three negative charges, it holds better and doesn't leach, but gets tied up, in many cases tied up beyond the plant's ability to use it. Green manure crops are useful for taking excess soil soup nutrients and putting them into their plant structure. Later when the green manure crop is worked into the soil, the nutrients are released slowly as biological decay occurs, continually refilling the soil's soup pot.

This soil soup needs to be more than just N-P-K if you are to grow healthy, high-yielding crops. Do a better job of adding a balance and provide soluble and slow-release nutrition. I find that using natural mined materials as my slow release puts the nutrients there over time. Natural materials contain many trace elements, in natural forms. Balance the materials used, split the applications, balance the soluble to slow release and stir the pot if necessary (tillage).

For the soil soup to give up its goodies, the "goodies" have to be there, but air and water also need to be in the pot. The plant has to have a large, active root system to extract from the pot. Organic farmers can only use slow release materials and no manufactured acidified solubilized materials. For the organic farmer to be successful and have a balance of highly exchangeable nutrients, he has to rely on soil organic matter cycles and biological activity. There is no law on the organic book that says you need to starve your crop. Organic farms can certainly yield as much as any other; the farmer just needs to be more skilled. He doesn't have the option of filling the pot with solutions of soluble nutrients. Nitrogen is a big one. He can grow his own; and many organic farmers' success is in direct proportion to how they accomplish this.

In the soil soup, different sources of materials used do react differently, and we must be careful not to add harsh salty materials. All sources of N-P-K (and S traces) are not the same, from the "soil soup" concept of viewing your farm.

There are things you can do when adding a crop fertilizer to actually alter what a plant receives and you can spread these over time. Use a balance of soluble to slow release, balance the nutrients, place them in concentrated zones near the crop roots and control the pH of materials used.

We manufacture a trace mineral mix with the trace minerals homogenized so each fertilizer pellet has a balance. This is a distribution advantage with more soil-root exposure. We also lower the pH of applied material. If a soil is near neutral pH and we add an acid pH (5.3 to 5.5) material around each fertilizer pellet, our exchangeability of nutrients lasts longer. They will not get tied up in the soil complexes so fast.

On our demonstration farm we look at rotations which include deep-rooted crops. Sweet clover is one crop we look at interseeding in a forage mix. The soil soup has holes in the bottom of the pot, and as water moves down in the soil, so do nutrients. Deep-rooted plants not only bring up some of these lost nutrients, but also leave root channels for future air and water exchange. Using plants to hold nutrients for the soup and keep releasing them is part of the value of interseeding and rye plow-downs, etc. These crops do more than provide nitrogen. To see research putting economic value on these practices solely based on nitrogen release is only telling a small part of the story.

It seems so simple and so logical. Why aren't more farmers fixing their soils? Soil fixing is an investment, like buying a piece of equipment. Once you have the soil fixed and the correction paid for, you can keep using it and get top performance by just taking care of it. The cost of the soil correction should not be added to that year's fertilizer program. It is paying for a lack of a balanced fertilizer program from the past. I always see the need for using crop fertilizers, because elements like nitrogen, sulfur and boron leach and need to be continually added. If the soil, with all its variations and capacities, provides a certain level of minerals, we can greatly add to this nutrient supply with a balanced crop fertilizer, properly placed for ideal recovery.

As the soils improve, you can reduce the crop fertilizer and get the same kind of crop you have been getting in terms of yield, but with a much higher quality. *Or*, you can keep the balanced row fertilizers at a higher rate, do a precision job of planting using higher seeding rates, and dramatically improve your yields. You have to earn the right to put more seeds out and increase yields. The limits and possibilities are beyond our present thinking.

Case Studies from Around the Country

I will now go through some examples of soils and areas of the country where we see different things. The examples and situations are from real farms. Different climates, different crops, different soils — these don't make any difference. The approach is the same: start with the evaluations, look for limiting factors and make the changes. Let's say you don't want or can't use green manure crops; that you don't want to rotate crops — it's just corn or corn and soybeans. How can this biological program benefit you? If you are using salt-type fertilizers or any kind of N-P-K blend, going to a complete, balanced fertilizer, properly placed, and doing the soil corrections will still give you results. It may not be the maximum potential your farm is capable of, but certainly an improvement.

As stated before, our soil correctives start with calcium and phosphorus. For the balanced crop fertilizer, select a low-salt fertilizer with a balance of soluble and slow release materials (Idaho phosphate, humates, slowly soluble ammonium sulfate, monoammonium phosphate, potassium sulfate, calcium sulfate and all the trace elements in sulfate form, except boron). The chelated trace element sources are good, but costly; they work best as a "pop-up" fertilizer or as foliars. Other things can be added, but only after all the basics are there.

If you have been using 200 pounds of 9-23-30, which is a mixture of DAP, diammonium phosphate (18-4-0), and KCl, potassium chloride (0-0-60), you are using a mixture that is too soluble and has extra ammonium from DAP and excess chloride from the 0-0-60. Besides the phosphorus and potassium, that's it. No sulfur, no calcium, no traces — and poor choices for the N-P-K. Switching to 200 pounds per acre of a 10-8-10-13S, with 5 percent calcium, 0.15 percent boron, 0.15 percent copper, 0.4 percent manganese and 0.5 percent zinc, plus added kelp, is a dramatic improvement. Note that the P and K levels are lower; you can only claim the soluble numbers on the label, so the slow-release phosphorus in the blend is not listed on the label. The solubility and numbers on the bag do not mean it is plant-useable. The nutrients can wash away, tie up, damage or inhibit root growth, interfere with other elements or slow down plant recovery. Soluble nutrients with high numbers and low price are the wrong reasons to select fertilizers.

If you have low soil phosphorus, you may choose a row fertilizer with a higher phosphorus level until the soil corrections are made. As stated many times throughout this book, it is not a numbers game. If you check under the specific crop section in the book, more details and reasons are given for the *whats* and *whys*.

The following examples are to give you ideas of how to approach fertilizer use and soil fertility. You need to look at your own operation to put together a working farm program that fits your needs and budget, and which addresses your limiting factors. No matter where you are, the principles are the same: mineralized soils, managing decay of residues, controlling air and water, and taking care of soil life. For your area you need to select farming tools, crops and fertilizers to fit the situation.

Example 1

Low CEC sandy soils from central Wisconsin and southwestern Michigan. Crops grown: corn-soybeans.

Test	Wisconsin	Michigan	Ideal for this soil
CEC	4.0	3.9	–
OM	1.4	1.4	2.0
H	0	2.1	0
P1 (ppm)	45	41	50
P2 (ppm)	49	44	100
% Base Saturation			
K	3.3%	3.7%	8.0%
Mg	28.7%	21.7%	25%
Ca	68.0%	53.6%	70%
ppm			
K	51	56	100
Mg	137	102	125
Ca	541	419	600
pH	6.8	5.7	6.8-7.0
ppm			
S	12	14	15
Zn	0.5	1.6	5.0
Mn	4	53	20
Fe	20	110	20
Cu	0.2	0.7	2.0
B	0.5	0.5	1.0

The ideal balance for this soil is impossible to achieve. Higher potassium levels are needed, and it is difficult to build and hold the sulfur and boron. These soils are formed with a mineral imbalance, and balanced crop fertilizer use is essential. Just sticking seeds in the ground on these types of soils will certainly return poor results. Neither of these farms uses irrigation, and in many years water can become a major limiting factor. Another real major limitation is the total mineral supply and the soil's ability to hold minerals. These soils can have a balance but will need a high level of crop fertilizers providing a balance of nutrients. These crop blends should contain both soluble nutrients for immediate use and a slower release material which will provide nutrients over time as the crop needs them. Timing and placement of fertilizers will also help ensure good results. Balance, concentration and recovery is what allows fertilizers to provide the crop's nutrient needs when it needs them.

For the Wisconsin farm, the soil is uniformly medium to low for most nutrients. Because the farmer grows row crops, the need to broadcast a soil corrective is not essential. It is essential to have a dry fertilizer delivery system at planting and also a liquid system for later nitrogen application. These light sandy soils are very difficult to farm organically, where no soluble fertilizers can be used. If you do want to farm them organically, you had better have a manure or compost source in your area or expect poor yields.

For the corn on this Wisconsin farm in a corn-bean rotation, a possible starter would contain:
- 50 pounds MAP
- 50 pounds natural rock phosphate
- 50 pounds granulated high-cal lime
- 50 pounds sol-po-mag or K-mag
- 50 pounds ammonium sulfate
- 50 pounds balanced trace mineral blend, providing at least 3 pounds zinc from zinc sulfate, 5 pounds manganese from manganese sulfate, one-half pound copper and one-half pound boron.

That is 300 pounds of starter, but you are not done. More nitrogen will be needed, and also more potassium. These are best applied when the corn is near knee-high. Another 100 pounds potassium sulfate and 30 gallons of 28 percent nitrogen would be minimal. In the liquid nitrogen I like to add humic acid. This is critical on those sandy soils to reduce nitrogen loss. I like placing this near the root zone, possibly dribbled on each side of the row and covered with soil by the cultivator. The potassium could be dry-applied in the root zone also. Because boron is so low and essential for sugar to move and calcium to do its job, in this case I would add another one-half pound of actual boron to the liquid nitrogen. This is the chemistry; we do not know the physical or biological properties of the soil.

Maybe rye or vetch was seeded in or after the previous soybean crop. This would help hold nutrients, provide food for soil microbes and allow for a reduction of fertilizer, especially nitrogen for now. After a few years at this fertility level, soils need to be retested and the crop blend altered to fit the soil changes.

Soybeans on this farm can also be fertilized with a dry row blend. If the soybeans are to be drilled, bulk spreading can be done, but rates should be increased by 25 percent or more. A possible blend for this farm would be:
- 50 pounds MAP
- 50 pounds K-mag or sul-po-mag
- 100 pounds potassium sulfate

- 50 pounds granulated calcium
- 50 pounds trace mineral blend, providing two pounds zinc, five pounds manganese, two pounds copper and one-half pound boron. Note the larger amount of copper. We have found this beneficial for healthy soybeans.

Again this is at 300 pounds/acre and is a fairly expensive blend. After three years reductions can be made. Retesting needs to be done. If this amount exceeds your budget, use the same proportions and reduce the application rate.

On this farm, with adequate moisture, corn yields exceed 150 bushels/acre and soybeans 50 bushels/acre. These kinds of crops can be grown even on these light soils. To help assure moisture, the soybeans are grown in rows like the corn, and after bean harvest subsoiling is done in the row down to 18 inches. The following spring corn is no-till planted in that zone. Corn roots like growing in decaying soybean roots, and with the deep subsoiling, the corn roots can go deep for moisture when in need.

The Michigan soil is included here to show a soil which needs a major correction. Note the pH, calcium level, hydrogen level, and manganese and iron levels. These all indicate the same problem — this soil needs lime. Because it also has a low phosphorus reserve level, adding a natural rock phosphate prior to liming will make the rock phosphate available for plant use faster due to the acid soil. I would start with 500 pounds/acre of rock phosphate correction in the spring, and in the fall add one-and-one-half tons of a high-calcium lime. This application may have to be repeated in three years, but retest first. Light soils change quickly with lower material applications, but need continuous crop fertilizer additions. The crop fertilizers for this soil would be similar to those for the Wisconsin soil except that adding extra calcium to the starters may be beneficial, and the manganese is not needed at this time. As the pH and calcium level comes up, the hydrogen on the soil test will go to 0, and manganese and iron levels will drop. There is no one exact way to correct these soils. For successful crops, minerals are necessary. Also any soil-building crops interseeded in growing corn or soybeans will be beneficial. If livestock manure or compost are available in the area, soil corrective additions and crop fertilizers can be reduced.

Example 2

Medium CECs (8-11) from Wisconsin, Michigan and Tennessee. The southern Wisconsin farm is an organic vegetable farm on which the farmer has been working toward soil balance for many years. The Michigan farm grows corn, soybeans, wheat and edible beans; it has no livestock and has been a heavy user of commercial N-P-K fertilizer. The Tennessee example is from a dairy farm which produces corn and forages in a rotation with controlled grazing.

This type of medium-range CEC soil is very typical for many of the farms we work with. Proper management, fertilizer use, crop rotations, cover crops and livestock manures create soils which are not only biologically active and include a lot of earthworms, but which have the potential to produce high-yielding, healthy crops with minimal input. I am only showing one soil sample from each farm; to put a farm program together, samples from all parts of the farm need to be analyzed. Soil correctives will probably vary with different farms, soil types and areas on the farms. Also, if livestock manure is available, it can be distributed to correct soil fertility and obtain uniform soils, crop fertilizers and feed supplies. On many farms, two or three crop fertilizers are blended for the whole farm, and soil corrections are varied along with application rates of the crop fertilizer selected to best fit the farm.

Test	Wisconsin	Michigan	Tennessee	Ideal for these soils
CEC	11.0	10.2	9.9	–
OM	2.8	2.7	2.8	3.0+
H	0	0	6	–
P1 (ppm)	90	46	74	50+
P2 (ppm)	130	74	103	100+
% Base Saturation				
K	3.0%	4.7%	5.2%	3.2%
Mg	26%	23%	26.37%	15%
Ca	70.9%	72.3%	62.6%	75%
ppm				
K	128	187	199	130
Mg	353	283	311	200
Ca	1,533	1,481	1,235	1,700
pH	7.0	6.8	6.6	6.8
ppm				
S	15	13	19	25+
Zn	6.3	1.9	1.7	5.0
Mn	25	16	78	20
Fe	29	59	43	20
Cu	2.4	0.9	1.0	2.0
B	1.1	1.1	0.6	2.0

Sticking seeds in the ground and harvesting these soils can produce good results. Adding a balanced fertilizer and having a biologically active soil with good structure gives excellent results.

On the Wisconsin farm, very little correction needs to be done. Magnesium levels can still come down, calcium can go up, and sulfur and boron will need to be continuously added. Sulfur and boron are anions, just like nitrogen, and need to be continuously in the fertilizer program. Testing and monitoring these elements assures you of not getting an excess. Most southern Wisconsin soils start out with magnesium levels in excess of 35 percent base saturation and calcium 55 to 60 percent. It is more difficult to reduce an element in the soil than to add it. Still, this Wisconsin farm is changing — calcium is coming up and magnesium going down. The objective was *not* to get an ideal-looking soil test, but to use the soil test to make a fertilizer buying decision. If you have high enough soil levels, don't buy more, and if they are low, add the mineral. Comfortable guidelines for many farms are for phosphorus addition (500 pounds natural mined phosphate) and/or 50 to 100 pounds MAP. There is no magic in that application rate, but with working on many farms over many years, the farms made progress toward balance and were happy with their crops.

For potassium I like to stay on the conservative side. I also like a blend of different potassium sources. Apply 200 to 300 pounds per year of sul-po-mag if both magnesium and potassium are needed. If only potassium is needed, use 200 pounds of potassium sulfate. I like making the potassium corrections in the fall. If soil potassium levels are low, select a crop fertilizer with higher potassium levels until the soil is corrected.

The calcium correction, if lime is needed, is usually in the one to two ton application range. If the soil pHs are high enough, kiln dust or gypsum can supply additional calcium in a more soluble form. For gypsum, 500 to 1000 pounds per acre is a good rate, and kiln dust is applied at 500 pounds to one ton per acre. Smaller amounts more often gives us good results and helps stay within the farm budget.

Soil trace element corrections can many times be made just using the crop fertilizer at high trace element rates. I like using the sulfate forms of trace elements. For a zinc correction, use three to five pounds of actual zinc or 10 to 15 pounds of zinc sulfate per acre. For manganese, five to 15 pounds of actual manganese can be applied. Copper can be applied at one to three pounds actual copper per year, and boron at one to two pounds actual boron. Very seldom do we need to add an iron correction. These are guidelines, and after a few years at these rates, do a retest and alter rates if necessary. At Midwestern Bio-Ag, we manufacture a trace mineral blend. We supply the trace elements in a rock phosphate and ammonium sulfate base, giving us a homogenized mix for better soil distribution.

Back to the Wisconsin soil test. This farm's only correction is 500 pounds gypsum, with one pound of boron added. These additions supply calcium, sulfur and boron — just what the soil needs. This is an organic vegetable farm growing a wide variety of crops. The crop fertilizer used is three tons of good compost per acre, a green manure crop, liquid fish and seaweed as needed. These additions supply extra nutrients above the soil's capacity to release them and also supply food for the soil life. The addition of materials like kelp and fish provides additional plant hormones, enzymes and many additional trace elements which are not tested for in the soil or in the crop, but which are needed in very small amounts. This farm grows healthy, high-yielding crops.

The Michigan farm is just getting started. The calcium:magnesium ratios are in fairly good ranges. You can also tell from the test that this farmer has been supplying the phosphorus and potassium with his present commercial fertilizer program. Enough is enough. These P and K levels do not need to be built any higher. The farm's limiting factors are the trace elements, soil structure and biological activity. For calcium and sulfur, 500 pounds of gypsum was applied, and the rest of the fertilizer was row-applied. A crop fertilizer providing extra zinc, manganese, copper and boron is being used. Clover is interseeded in the wheat along with the spring nitrogen. Grain rye is fall-seeded following soybeans. The corn crop is interseeded with a clover-rye grass blend at last cultivation when additional nitrogen is side dressed. The farm has cut herbicide rates and also has reduced nitrogen applications by half. Using nitrogen sources with added sulfur and properly placing them at the proper time, along with the green manure crops, has allowed the farmer to reduce commercial nitrogen purchases. Crop rotations have been tightened; no corn-on-corn is grown, and insecticides are no longer used. Soil health, crop yields and crop quality have all improved.

In the test from Tennessee, P and K are also high. First things first, and if the P, Ca and pH are low, those are the first corrections. This is a dairy grazing farm. They buy some feed, have a lot of cattle and will have to deal with excesses more than deficiencies. P and K are high and there is a lot of manure, but this does not mean no fertilizer is needed. To balance this soil, it will take more zinc and calcium because of the high P and K. Also note that even with a lot of dairy manure, calcium, sulfur, zinc, copper and boron are still low. On many dairy farms this is true and the same goes for gardens where a lot of compost and manure are used. Ninety percent of the gardens I have treated are low in calcium, sulfur and boron. Sulfur and boron being anions like nitrogen, leach and continuously need to be added to the soil.

To obtain feed quality and keep insect pressure low, these high potassium soils, with probably excess nitrogen from all the manure, are a challenge. I'm worried about nitrates or poor quality protein in the forage from the extra nitrogen. Sulfur is absolutely essential here for building these proteins.

Palatability is another concern. Cattle don't like to eat around their manure piles and now the whole field is like that. They don't like feeds high in potassium and nitrogen. It is always easier to add something to a soil than to try to take it away. The high potassium and nitrogen give rapid growth. High potassium means stronger stems, but coarse and hollow stems, low in energy. The forages need to be harvested early, which also means lower sugar and energy. That's why dairy farmers feed a lot of starch and fats to their cows to get production. The way they fertilize and the soil balance gives them low-energy, fibrous feeds.

For fertilizer correction to this Tennessee farm, calcium needs to be applied because it is grazing land and the pH is already 6.6. Lime would be a poor choice. It does not move in the soil much, it is too coarse and it is not very soluble. Kiln dust would work very well here. It is extremely finely ground, and may have sulfur present. Because of the chemical complex of calcium after going through the burners, kiln dust's calcium is in a highly soluble state. Results are rapid even when surface-applied. One thousand pounds per acre would be a good rate for this farm.

In addition to calcium, trace elements and nitrogen are needed. This is a grass farm. Establishing more legumes in the pasture may reduce nitrogen rates, but for now more nitrogen will be needed, even with all the manure used. These farms typically apply 40 units of nitrogen after each grazing or cutting. Many of them get five cuttings. This gives you that big pile. Also, being in a higher rainfall area and trying to grow quality forages, frequent fertilizer applications work best. I would use 75 pounds of ammonium sulfate, which would provide both nitrogen and sulfur. To that I would add one pound of zinc, one-fourth pound of copper and one-half pound of boron. This will be applied three to four times during the growing season.

A final note: notice the manganese-to-iron ratio. Soils with higher pHs that have higher manganese still indicate good biological activity. During a hot and possibly dry summer, this same soil may only test one-half that manganese level. This is the only clue on a soil test about biological activity. These fields are not tilled and have constant supplies of manure added. Both conditions promote biological activity.

Example 3

High CECs (12-25). From Wisconsin and Illinois.

Test	Wisconsin	Illinois
CEC	13.5	22.9
OM	3.2	4.4
H	0	19
P1 (ppm)	47	34
P2 (ppm)	56	58
% Base Saturation		
K	4.4%	2.2%
Mg	36.8%	29.7%
Ca	58.9%	49.1%
ppm		
K	229	194
Mg	595	815
Ca	1,588	2,246
pH	6.8	5.8
S	17	14
Zn	2.3	3.9
Mn	9	15
Fe	59	165
Cu	1.0	2.1
B	1.3	0.08

Here I have used only two examples. You can find many more typical examples. Farm fertilizer programs need to be put together for individual farms. You can tell by looking at the soil test how the farmer is farming. In the Wisconsin sample, note the high magnesium level along with the high P and K. This is a crop and beef cattle farm. The farmer told the fertilizer supplier he didn't care about quality; he had lots of cattle to feed and wanted tons. The farmer was sold nitrogen and potassium and applied them at high levels.

With this heavy soil and high magnesium, if your objective is to just change the Ca:Mg ratio, it will be a long uphill battle. The dolomite lime quarry is near this farm. Magnesium is high; certainly don't add more. Adding high-calcium lime to soils with this high pH won't change much. Adding more potassium or magnesium at this time would certainly not give much of a return on investment, it might even give a negative return.

This farmer needs calcium. Here again a gypsum or kiln dust source is much better. The pH is high, so adding calcium with sulfur works the best. Due to the high CEC, these soils have potential to hold more nutrients, and if you want to change them, you have to add more fertilizers and liming materials. If this soil has the mineral balance, good biological activity and soil structure, it will have the potential for a 200-bushel corn crop with little more than sticking seeds in the ground and adding a little nitrogen. If you look at the corn yield averages for Illinois, you will see that that is not what they are harvesting. Many heavy soils have their limitations — physical, biological and chemical. They also have a lot of potential.

For the Wisconsin farm we put on 1,000 pounds BioCal (a calcium, sulfur and boron source) on all soybean and forage fields, and the crop fertilizer was MAP (50 pounds) along with ammonium sulfate (100 to 200 pounds) with two pounds zinc, five pounds manganese, one pound copper and one pound boron added. Modifications were made for specific crops, and additional liquid nitrogen was side dressed for corn at 20 gallons per acre. Crops have greatly improved, but to change these soils will take time.

For the Illinois farm, there is not the very high CEC. It has lots of potential, but is a farm with poor crops. This farm was only using nitrogen in the anhydrous form at 200 pounds actual as the corn fertilizer and nothing for soybeans. That high excess nitrogen surely did contribute to the low calcium and low pH. What do you think the soil structure and biological activity looks like? Hard, tight ground with not a worm in sight. Do you believe switching to no-till will solve this farmer's problems?

You need to start with lime, lots of lime, and work it in those top six to eight inches. High calcium lime fits this farm, and it may take a total of eight tons over time to fix the soils. Nitrogen rates need to be reduced to avoid calcium leaching. This is the kind of land that the term "farmed out" came from. Subsoil this land and get green growing crops. Interseed the corn, band the herbicides, side dress nitrogen, and rye or vetch can be planted late in beans or after harvest. The fertility program can use some phosphorus, row-applied at planting. Some potassium is also needed, along with sulfur, zinc, manganese and boron. With these low pH, high-ammonia soils, iron is usually high. We want a 1:1 manganese-to-iron ratio, but it is now 10:1. Applying the lime and waiting a few years should improve the situation. It will take time and money to fix this land, and it is rented. You may want to let someone else try to grow profitable crops on it. It's sad to say, but a lot of land in the Upper Midwest commands high rents and is in this shape. The farmer farms lots of acres trying to make a living. The shape this field is in means yields of one-half the soil's potential. You hear many comments that if we all go organic, we will need twice the acres to grow the crops. I'm sure if you "fixed" this soil and farmed it organically, professionally providing good management and mineral balance, it could far out- yield what is being produced now. It would be greater quality and certainly better for the environment.

14

BIOLOGICAL FARM CASE STUDIES

    This chapter is about real farm examples, with their stories of actual changes made and results obtained. The examples were provided by Midwestern Bio-Ag consultants and their farmers. I will give a description of their operations, when they started changes and how. I will also discuss observations and future changes planned. These are not examples to copy, but just to provide ideas for putting together working farm programs. If you have a plan and know where you are going, you will probably get there. On the other hand, if you keep doing what you have been doing, you will keep getting what you have been getting. These examples are from all over the country, with many different kinds of farms, all trying to achieve the same results: healthy soils, healthy crops and healthy livestock.

    In the early 1940s, Louis Bromfield was working his Ohio Malabar farm, observing and writing about biological farming. In his book, *Return to Pleasant Valley*, a chapter called "Why Didn't Anyone Tell Us?" is most interesting. Bromfield comments:

> *All too often the farmer is told to lime his soil without being told the faintest thing about what the effects of liming are beyond the fact that if he limes he can raise legumes. He is not told of its effects upon ionization processes related to iron and aluminum or the wonderful catalytic properties*

of limestone. He is considered too stupid and uneducated to understand about trace elements. He is treated, all too often, as if his only function was to take the advice given him without question and simply go ahead grate- fully and do as he is told. As one of the countless farmers who visited Malabar put it, 'We like to visit successful farms. Looking at research and experiment stations gives us ideas, but nobody puts it together in a working pattern for us.'

An experiment plot is not the same as a farm. What works on a prepared plot does not work in a field with two or three kinds of soil. At a success- ful, well managed farm we can get an immense amount of information and a lot of good ideas in just a few hours.

Bromfield further comments:

Somewhere along the line there is a missing element in our education, and I suspect it is the pilot farm. A farm operated by a legitimate farmer himself in terms of the average farm. A farm where visiting farmers may come and go over the place field by field and animal by animal, knowing that the whole place is not operated on taxpayers' money but profitably on the farmer's own income and capital. It would be a place where he could examine the texture of the soil and the health and vigor of the livestock and look over the fence and know he is seeing a productive and profitable field, and that he can find out how it was made that way. We should remember that every farmer believes what he sees.

I believe he was absolutely right and it is equally true today as it was back then. Biological farming will become the "conventional" way in time, as farmers see and learn from successful operating farms. This is my next big project, to put together a directory of model, pilot, successful bio- logical farms. Farms where the farmer is willing to share his experiences, both the successes and failures. This chapter is a start on that list.

Midwestern Bio-Ag has its own such farm for demonstration and learn- ing. Each year during the third week of August we have open house, invit- ing farmers and educators from all over the country. If you or someone you know who is a successful biological farmer would like to be included in our model farm list, let me know.

Model Farm No. 1

Kalmes Farms, Inc., Rollingstone, Minnesota

Kalmes Farms were established in 1951 when Eugene Kalmes pur- chased a 340-tillable acre dairy farm near Rollingstone in southeastern Minnesota. Right away he put into effect conservation practices which included contour strips, waterways, ponds and reduced tillage. Eugene implemented a five-year rotation which is still used on the farm: corn, corn, hay, hay, and then a new seeding mix of oats and peas. Manure from the dairy was surface-applied daily. Plowing was done on sod fields going to corn. All other ground was chisel-plowed in the fall. A spring-toothed cultivator followed in the spring.

Crop fertilization on corn was done in-row with 100 pounds of 9-23-30 and 100 units of ammonia. This fertilization program was utilized until the mid-'70s. In the late '70s, they switched to 60 gallons per acre of 28 percent nitrogen with a post-plant herbicide that included one quart Atrazine and two quarts Lasso as a post-plant weed and feed mix. Insecticides used at that time were Counter and Lorsban.

Alfalfa fertilization used 200 pounds of 0-0-60 per acre on all hay acres in the fall and two tons of dolomite lime from local quarries on new seedings. Eugene and son Bob commented that they were never very satisfied with stands and yields in those years.

In the early '80s, they used a liquid lime product, calcium oxide, blended with alfalfa seed and sprayed on at 1000 pounds/acre. They were well pleased as stand vigor increased, but the supplier lasted only three years and the product was no longer available.

Soil testing in July 1989 showed fields that averaged 62 percent base saturation calcium and 28 percent magnesium, an inferior 2:1 Ca:Mg ratio. Phosphorus and potassium levels were medium to low, and trace elements were all low.

Consulting during this period was provided by the Land Stewardship Farming Program, which included on-farm research in the early '80s and demonstrated sustainable farming methods. Bob saw on his test plots that purchasing additional nitrogen on corn was not profitable, as there were no significant yield increases. "We had a lot of really good and interesting meetings to help us see our way," said Bob. Gary Zimmer was one of the speakers on biological farming who made sense to Bob. "Dad still cannot believe we can grow corn without purchased starter, nitrogen or insecticides," he said. Now after ten years, Bob sees yields that are satisfactory compared to neighbors in the area.

Also at this time, plots were used to evaluate the amount of nitrogen needed to grow a crop of corn when using different rates of manure application. To this day, Bob uses these numbers for his manure application rates. They are: sod to corn, 10 tons liquid manure per acre in the fall; corn on corn, 20 tons; and corn to beans, 10 tons. Applying liquid manure to hay fields is avoided. Also at this time, Bob started using high-calcium lime (35 percent Ca, less than 1 percent Mg) from quarries in the Minnesota/Iowa border area. This was applied at two tons per acre on new seeding ground every fifth year, costing approximately $25.00 per acre applied. Consulting staff from the Land Stewardship Project interpreted and did recommendations from Bob's soil tests, using the Albrecht theory of soil balance.

Farming with the Midwestern Bio-Ag program began in 1991. Bob and his wife Betty joined the farm in 1972, and the "retirement" of Eugene in 1986 increased their responsibility. Mike Lovlien, Bob's Bio-Ag consultant, advised him that he should continue the use of high-calcium lime and BioCal, and use soil testing to better manage the placement of manure by identifying fields that tested low in potassium and phosphorus.

Currently the soils on the Kalmes Farm have a base saturation average of 68 percent calcium, 3.8 percent potassium and 27 percent magnesium, with 58 ppm of P1 phosphorus, 14.5 ppm of sulfur and average to low trace elements.

Test	1989	1997
CEC	10.95	10.95
OM	2.0	2.47
P1 (ppm)	32.5	58.75
P2 (ppm)	40.5	76.5
% Base Saturation		
K	2.9%	3.8%
Mg	28%	27.3%
Ca	62%	67.3%
S	13.5%	14.5%
ppm		
Zn	2.0	3.0
Mn*	45	19.25
Fe*	49.25	33
Cu	0.95	1.12
B	0.8	0.7

(1989 samples were from soils with corn going to new seeding;
1997 samples were from soils with soybeans going to new seeding)
* Manganese and iron samples are two times higher in 1989 than in 1997,
which may suggest differences in mineral uptake from the soybeans rather than the corn.

Each year Bob has been doing soil corrections on new seeding ground, which occurs once every five years. Recommendations are based on soil correction application. The 1996 soil sample results indicated that the sampled fields needed a correction with a custom mix: 50 pounds 0-0-50, 125 pounds cal-sul, 75 pounds Bio-Ag Trace Minerals 5-5-5 (Wee Mix), which was applied prior to seeding in the spring. This was a $29.00 per acre cost for the custom blend. Also applied were 1000 pounds BioCal per acre, costing $45.00 per acre applied. The total cost per acre was $74.00. On the 120 acres of forage ground, a continuing crop fertility program consists of 500 pounds BioCal at $21.00 per acre and 200 pounds of Bio-Ag's 4-15-15 high-trace fertilizer blend at $36.00 per acre, with a total cost of $51.00 per acre.

Mike continues to be impressed with Bob's focus and commitment to a quality forage program, apportioning a majority of his crop input costs to his quality forage program and the remaining dollars going to crop expenses.

Bob's corn enterprise since 1989 has been quite impressive. He has been able to reduce purchased nitrogen and in-row starter to zero. Bob uses 12 pounds of Bio-Ag's Bio-Root at a cost of $10.00 per acre, applied on second-year corn only. There is tremendous root mass where it is applied. He continues to cultivate, usually in June. Herbicide applications are on corn following sod, at 1.5 quarts Atrazine and 8 ounces Amine, with a cost per acre of $6.00. The second-year corn receives 2 quarts of Lasso and 3.5 pints of Marksman, costing $24.00 per acre. His 1997 crop averaged 170 bushels per acre dry basis. Of the 100 acres planted, 30 acres are harvested as corn silage and 70 acres as high-moisture grain. This year Bob did research into seed corn varieties developed for quality corn silage and feed ability. Results are not yet available.

Bob's soybean enterprise consists of 50 acres, with inputs mainly manure and spray: 12 ounces of Pinnacle and 1.5 gallons of Pursuit, costing $37.80 per acre applied. Bob drills his beans and had an exceptional year in 1997, with yields averaging 60 bushels per acre. The soybeans are all on-farm roasted and stored. Soybeans were added in 1996 because in the past all off-farm sales of small grain forages were sold in years of excess with little return over fixed cost. Now, due to the loss of 50 acres of corn to soybeans, Bob is buying 6,000 bushels of corn. The quality differential in corn and mineral uptake does not vary widely, whereas homegrown protein is much more valuable in Bob's dairy nutrition program. Prior to 1996, Bob's protein bill was averaging $30,000.00 per year, with questionable quality, as every batch of purchased pellets would look different. Currently, the corn is costing $15,000.00 to purchase, netting Bob's bottom line $15,000.00. Production costs per acre between corn and beans are similar, so a net profit can be realized.

On the dairy, with the tremendous yields of soybeans in 1997, Bob is able to feed all homegrown protein at 4.5 pounds per cow per day for 110 cows all year. From the nutritional standpoint, this is a great asset to the well-being and health of the herd.

Alfalfa quality is very impressive on the Kalmes farm. Mineral uptake in the plant is very good. The 1997 first-crop feed test results were: 1.64 percent calcium, 0.44 percent phosphorus, 0.35 percent magnesium, 2.81 percent potassium, 0.35 percent sulfur, 24 percent crude protein and a nitrogen:sulfur ratio of 11:1, which indicates a greater amino acid balance. The fiber levels were low at 32.4 percent A.D.F. and 37.1 percent N.D.F. Bob harvests his forages rapidly, usually within two to three days, making his quality crop even better. Inoculants are used to burn and consume oxygen in the 25-by-90-foot Harvestore converted to top-unloading. The silo has rapid cool-down in three to four days after filling.

Dry cow grass production is a special program on the Kalmes Farms. A balanced dry cow program is essential for optimum dairy cow lactations. Midwestern Bio-Ag believes it all starts in the soil, where anionic feeds are produced. Since 1995, Bob has applied no manure or calcium to the 20 acres that comprise the program. What is applied is Bio-Ag's anionic fertilizer blend of 14-15-0-13S at 200 pounds/acre at a cost of $31.00 per acre. The hay is then harvested in large square bales that are to be fed only to dry cows and springers. This management plan has really helped to get fresh cows on track and milking.

The dairy cow feed ration consists of high quality alfalfa forage with the same high quality mentioned above. At present, the cow group is receiving (all amounts on a dry matter basis): 22.7 pounds haylage, 9.5 pounds corn silage, 11.8 pounds shell corn, 4.5 pounds cottonseed, 2.8 pounds roasted soybeans, 1.7 pounds hay, eight ounces of Bio-Ga's TMR base mix, two ounces of calcium carbonate and two ounces of salt. Bob free-choices Bio-Ag's Grazier Salt/Kelp Blend at about three bags per week. Ration totals are .76 NEL 17 percent available crude protein, 53 pounds dry matter intake, 85 pounds milk, 62 percent forage-to-concentrate ratio.

Production results: RHA

June 1995	22,096 pounds	816 Fat 184 days in milk
January 1997	24,094 pounds	885 Fat 172 days in milk

Some of the farm improvements that have occurred since the addition of Midwestern Bio-Ag's program are:

• Doubled the capacity of the liquid manure pit so it will store for 12 months. This will be a great asset to soil life, nutrient recovery and recycling.

- Increased the soybean acres so that no outside protein purchases will be needed.
- Alfalfa yields have been up from prior years, which will aid in promoting higher forage feed rations.
- A new heifer barn facility was completed in the fall of 1997, which will aid in faster gains, enabling better management for all age groups.
- Recent soil tests show all fields beginning to average out. Nutrient levels are more evenly distributed across all fields and farms. This allows for easier feed ration balancing, as quality does not vary greatly from field to field.
- 1997 was the first year the Kalmes Farms ever sold 18 springers; a real plus for the bottom line.
- 1997 was the first year the hoof trimmer noted the hardness and quality of the hooves, not "soft and mealy" as in the past. "I can't believe it's the same herd" was a comment at a recent hoof-trimming session. Limiting factors — things to improve:
- Heifer management and nutrition.
- More emphasis on proper soil conditions before tillage and planting so that it won't set up for poor root health and air/water movement; i.e., less weed pressure and deeper crop root penetration.
- Would like to go to a higher forage ration for the dairy cows and still hold production, reproduction and profitability.
- Manure management could use Bio-Ag's Idaphos to make a good material even better.
- The use of fall cover crops will be a next step as technology improves.

Bob is a great example of a biological farmer. He is open-minded. When something makes logical, common sense, he is willing to do whatever it takes to make it work. He understands soil health and can identify its merits. He is willing to think and work "outside the box." Bob is in tune with the signs of optimum herd health and can identify change that is required of a good herdsman.

This farm story was written by Mike Lovlien, Bio-Ag consultant in southeastern Minnesota.

Model Farm No. 2
Pat Leonard Farm, Blanchardville, Wisconsin

This is a conventional dairy farm located in southwestern Wisconsin. Before going on Midwestern Bio-Ag's program in 1993, Pat used a rotation of corn-corn-new seeding (with either oats or barley nurse crop) — hay-hay-hay (or sometimes one more year hay). He applied manure to the corn ground and used commercial salt fertilizers. His corn starter was 200 pounds/acre of 9-23-30. Alfalfa was top dressed primarily with 200 to 500 pounds/acre of a potash blend (0-14-42) with sulfur and boron. Nitrogen was obtained from plowdown sod and manure, and no additional nitrogen was applied the two years before 1993. According to soil pH, no lime was needed for several years before 1993.

For weed control, Pat used Atrazine in the past, then switched to a Bladex + Banvel mix. Round-Up was used on sod in the fall. Insecticide was used on second-year corn.

No yield measurements or program costs are available from Pat's previous practices. Pat was basically a good manager, but was trying to improve in several areas: feed quality, mineral balance and soil health.

After going on Bio-Ag's program, Pat grew mostly the same crops, but added some soybeans to supply his herd homegrown quality protein.

The manure management practices stayed the same; however, in 1995, Pat began adding Ida-Phos to the gutter.

For corn starter, Pat uses Bio-Ag's 10-9-10, and for alfalfa Bio-Ag's balanced hay fertilizer at 200 to 400 pounds/acre. He applies 700 to 1,000 pounds/acre of BioCal on oats/seeding and all hay acres. No additional nitrogen is used, with sod plowdown and manure supplying crop needs. No lime has been used, as indicated by soil pH; BioCal supplies needed calcium.

For weed control, Pat uses Round-Up on sod to control quack grass, with other herbicides being used after corn is planted. 1997 was the first year no insecticides were used. Pat was having problems with root damage and corn going down when he did use corn insecticides. In 1997, he used Bio-Root on all second-year corn with excellent results; the corn had noticeably larger roots and stood well. There was rootworm damage on a small acreage of corn that had insecticide applied along with a conventional salt-based fertilizer as a demonstration plot, with the local Co-Op comparing their program with Bio-Ag's program.

The total cost of the program is not available, but is about the same as Pat's past program on hay and a little higher for corn. Corn yields are as high as or better than on his past program, and hay yields are about the same as in the past. Mineral levels in the hay are good — at or near target levels. Since the hay is at its peak for a longer time, it is easier to get high protein. Pat says his corn looked healthier and seemed to withstand more stress than neighbor's corn.

For the future, Pat would like to eliminate the use and cost of all herbicides.

This information was supplied by Bio-Ag consultant Duane Siegenthaler, New Glarus, Wisconsin, who consults in southern Wisconsin and Indiana.

Model Farm No. 3

Mike Dietmeier, Orangeville, Illinois

Operating a corn-bean-wheat-beef farm in northern Illinois, Mike has been working toward a biological farm for a long time. Duane Siegenthaler is his consultant. Mike supplied the following write-up.

Back in the early '80s when we were still dairying, we used 0-0-60, 0-46-0 and 28 percent nitrogen with the herbicides. By the late '80s I switched to a liquid starter program. In some years I would broadcast 0-0-60 and ammonium sulfate mixed. At first I used 10-34-0, then in later years 9-18-9. In 1986 I first started side dressing 28 percent with the cultivator.

During those years, I was getting tired of poor animal health, weeds and toxic chemicals. I tried cutting herbicide and nitrogen rates in half on rented land, but that was not very successful.

In January 1990, we sold the cows and in 1991 started on Midwestern Bio-Ag's dry starter fertilizer.

We now feed out about 100 to 180 head of feeder cattle per year, and they do not produce nearly as much manure as the dairy cows. I clean the shed and lot about once a month, and that only covers 20 to 25 acres a year.

In 1997 I had only 20 acres of corn-on-corn out of the 305 acres of corn and 280 acres of soybeans. It seems like the only way to get really good yields of corn is to rotate. Now with a newer combine and head, that many beans aren't so bad. Corn yields were from 120 to 220 bushels/acre, with 160 average. Beans yielded 35 to 65 bushels/acre, with one yield check of 80.26. I would not put wheat after corn any more. There is too much residue, and no-till wheat after beans seems to get better stands, which equals yield. I haven't noticed much change in test weights. This year (1997) corn was between 54 to 55 pounds and beans were 57 pounds The wheat hasn't had a good test weight for a few years now. Sometimes I get a little discouraged with it and think I won't grow it any more, but fall comes and I plant it again. This fall it was 30 acres.

For cover crops I buy rye and spread it after corn and before beans at a rate of one to 1.5 bushels per acre. Also after wheat I sometimes plant oats at about 1.5 bushels before corn the next year.

Right now my fertilizer program is, for corn, 230 pounds of 9-9-9-10S plus traces with the planter and 20 gallons of 28 percent with thio-sul side dressed at cultivation. For soybeans, I apply 700 pounds of BioCal and 100 to 150 pounds of 18-6-1-19S plus traces broadcast, because I am drilling the beans. For wheat I am also using 700 pounds BioCal and 250 pounds of 28 percent. For hay, it is 700 pounds BioCal and 300 pounds of the fertilizer Duane recommends.

I have been using a half-rate of herbicides and getting good control — some years better than others, so I guess I haven't felt secure enough to cut it more so far. I did buy a twelve-row rotary hoe in 1996 and used it on about 150 acres out of 450 acres and saw good results. Then in 1997 I used it on all the corn and again saw good results. I may experiment and cut herbicide a little more next year if conditions are right to rotary hoe. I hoed a 70-acre field in about one hour and forty minutes. This year (1997) I was spraying beans. I didn't want to clean the sprayer out until I was done with the beans, so every day I hoed about 100 acres or so, and it really knocked the weeds back. With the beans being drilled, I feel a little uncomfortable cutting herbicide more than half. Sometimes in problem areas I don't cut at all. I use mostly STS beans, and the cost of the full-rate program is $18.00 per acre. In past years when the co-op sprayed my beans, it's been in the $40.00 and up range, with less yield.

I would say that probably low calcium and phosphorus are my limiting factors. The last two years I have applied a lot more BioCal than before. I like the things I see in the soil's tilth and the good, earthy smell, plus I feel good about the things I am doing and the way I am doing them.

Model Farm No. 4

Steve Hooley, Lagrange, Indiana

This is one of those real successful grazing farms, located in northern Indiana. Steve has nice cattle, good production and a wide variety of forages. A program for managing soil fertility and growing quality feeds is at work. Success is doing a lot of things right, and this farm does that. Steve's consultant is Duane Siegenthaler.

Before starting on Midwestern Bio-Ag's program (1995), Steve's operation was rotational grazing of dairy cows and heifers during the summer, with the animals spending the winter mainly in free stalls and a cement lot. Besides pasture, he grew mostly hay and haylage, with a little corn for silage. Steve's rotation was hay-pasture-corn. He tried to cover all acres with winter manure with a side-delivery spreader. For fertilizer, Steve used conventional liquid and dry products, including 28 percent on corn and legumes in pasture and some urea on pasture, and he used high calcium lime at one ton per acre every other year (to maintain a pH of 6 to 7). Boron was applied to hay fields along with liquid fertilizer. Soil tests were made by the local fertilizer company, mainly for N-P-K. He controlled weeds in corn with herbicide and in pastures by mowing. Insecticide was used in hay fields. Steve's costs were normal for hay and corn, with fertilizer running $30.00 per acre and herbicide $15.00 to $20.00. Typical yields were 3.5 to 4.5 tons per acre of hay, 3.5 tons (dry matter basis) for pasture and 125 bushels per acre of corn. Milk production ran 6,130 to 6,500 pounds/acre. He seldom did feed quality tests or test weights. Steve was dissatisfied with the way things were going, saying that he was not seeing the uniform results or production in pastures that he wanted, even with applying high-calcium lime.

On the Bio-Ag program, Steve is working toward a seasonal controlled grazing system (dairy). Besides pasture, hay and haylage, he is growing annual pearl millet. Other than what the cows deposit while grazing, manure is spread on acres not heavily grazed. For fertilizer and soil correction, Steve now applies BioCal every year (amount indicated by the soil tests) and 150 to 350 pounds of trace blend fertilizer, depending on soil tests. For nitrogen, pasture legumes supply much, but a little ammonium sulfate is added to the fertilizer blend. Steve's new crop rotation is hay (one year) — pasture (four to five years) — annual crop (one year).

Steve has dropped herbicides and relies only on mowing pastures. No insecticides are used on pasture or hay fields, and he is going to try ladybugs for leafhopper control. Current costs per acre are $60.00 for fertilizer and about $100.00 to establish pasture (seed plus fertilizer). Yields have been better, with 3.5 to 6.0 tons per acre (average 4.5 to 5.0) on dry matter basis, and milk production is up to 7,500 to 8,000 pounds per acre. Steve says feed quality is up and there is more of it. Cattle are using less and less free choice mineral, breeding is better, with a 67 percent first-time conception rate, and vet bills are only $25.00 per cow and calf. Besides free choice mineral, Steve uses salt/kelp and buffer in his rations. Overall, he is getting 1,500 to 2,000 pounds more milk per acre on 30 to 50 percent less grain plus supplements, which equals more profit.

In comparing soil samples from 1994 and 1997, phosphorus and potassium have remained about the same, with a wide range of potassium (80 to 200 ppm). CECs are from 6 to 12, with some low areas 20 to 30. The calcium base saturations were 63 to 73 percent in 1994 and now are 70 to 80 percent.

Steve's limiting factors include the need for some tiling (some soil currently needs to be aerated), and the distance that products have to be shipped.

Model Farm No. 5
Dave Campbell, Maple Park, Illinois

This is an organic crop farm, begun in 1988, raising soybeans, hay and grain on 153 tillable acres in northern Illinois. Previously the land had been rented for 30 years, with corn and soybeans raised. The soil had poor tilth and high erosion, even on flat bottoms, and weeds such as sunflowers and jimson weed were prevalent. Dave's consultant is Keith Ostby of Monroe, Wisconsin, who provided this information.

Dave started out in 1988 by growing 90 acres of hay. He grew some corn, but it was very weedy, averaging only 30 bushels per acre (the neighbor who combined it felt so sorry for him that he didn't even charge for combining). To help cash flow over the years, Dave sold hay to horse owners. Each year he would plow up one-third of the hay ground. He has grown many cover crops, including rye, clover, alfalfa, oats and vetch.

By 1991, Dave was certified organic on all the farm. The last four years he averaged about 40 to 43 bushels per acre of soybeans. In 1994, beans averaged 50 bushels.

Dave started out with very low phosphorus and potassium levels, and calcium was low also. P1 was around four to five, and potassium in the 60s. He put on high-calcium lime and has been using 200 to 350 pounds/acre of fertilizer. There has been a positive move toward better soil fertility. In 1995, Dave started using BioCal. His hay fertilizer is broadcast, and on soybeans, application is over the row, with gypsum or high calcium lime used in the insecticide boxes. Tillage has been plowing, disking or field cultivating in the past. Now he uses rotavator and chisel for most tillage and field cultivator for second passes.

Today, the soil structure is loose and some earthworms are appearing. One of Dave's horse hay customers told him his vet could not believe how little grain he was feeding his horses, but how good their condition was, which reflects on the results of Dave's hay program.

Model Farm No. 6

Randy Strey, Hortonville, Wisconsin

This is a hay, corn, soybean, wheat and cash crop dairy farm in northeastern Wisconsin. There had been a stray voltage problem which has been corrected, so milk production is ready to climb from better soils and crops. Randy's consultant is Clem Greisbach of Appleton, Wisconsin, who supplied this information.

Randy started the Midwestern Bio-Ag program in the late '80s or early '90s. His crop rotation is still corn-soybeans-hay, but it has been tightened, with only one to two years of corn (previously two to four) and three years of hay (was four to five). Randy used to use manure only on corn ground; now he also puts some on beans, but uses smaller amounts, spread over more acres. Before going on the Bio-Ag program, Randy used a moldboard plow on about 205 of acres and chisel plow on the rest, plus field cultivator; now he uses DMI and field cultivator, with no-till new seeding on wheat stubble.

Randy used to fertilize with 200 pounds/acre of 9-23-30 plus 100 to 160 units of anhydrous ammonia on corn and 200 pounds/acre of 0-0-60 on hay. He did not use any lime. Now, using Bio-Ag's fertilizers, he uses 150 to 200 pounds/acre and 30 to 60 units of 28 percent on corn, 150-200 pounds/acre of 4-15-5 blend followed by 500 to 700 pounds/acre BioCal on hay ground, and 1,000 pounds/acre BioCal for soybeans.

The soil audits from 1993 and 1997 show some changes. The P1 phosphorus has risen from 32 ppm in 1993 to 62 in 1997, potassium has dropped from 205 ppm (very high) to 150 (near the target of 143). Base saturation of calcium has risen from 59.8 percent to 63.6, and sulfur has increased from 16 ppm to 22. Zinc and boron levels have risen, and manganese has dropped (it was very high).

Randy used to use full-rate herbicides plus cultivation for weed control; now he uses three-quarter rate herbicides, cultivation and some rotary hoeing. He has dropped insecticides.

So far, Randy hasn't seen any yield differences, but has noticed that his previously coarse-stemmed hay now has fine, better quality stems. A 1997 feed test of first-crop haylage showed very good figures, such as 0.65 NEL, 1.71 percent calcium, 0.40 percent phosphorus and 0.35 percent sulfur. His soil is looser, with water soaking in faster after a rain. There are more earthworms. For the future, Randy wants to get higher yields and better feed quality.

Model Farm No. 7

Marvin Hill, Ruth, Michigan

Marvin's Bio-Ag consultant is Pete Creguer from Minden City, Michigan. Marvin started working with Midwestern Bio-Ag in the late 1980's. He has a small dairy herd and grows cash crops on 350 acres. Crops include edible beans, wheat, barley, corn, alfalfa, soybeans and rye for seed.

Prior to working with Bio-Ag, anhydrous ammonia was applied as the nitrogen sources at 150 units. Potassium chloride was bulk spread on hay ground and commercial 9-23-30 dry starter was row applied for corn and beans. Tillage was fall mold board plowing with discing and field cultivating in the spring.

The soils are high in calcium and have high pH's with medium fertility levels. Ammonium sulfate and gypsum for added sulfur and soluble calcium fit this farm well. Due to the soil's high pH, using a fertilizer with a low pH does provide greater fertilizer efficiencies. Livestock manure continues to be used to correct the fields with the lowest fertility. In addition, manure is lightly covered over the remaining acres, which is wise practice for any farm.

With all the crops grown, the rotation was good except for hay which was left in for 5 or 6 years and sometimes corn was grown on corn. Now, the hay fields are left for just 3 years and no corn on corn is grown. Changing to more cover crops was another improvement. Clover is now interseeded in the wheat in the spring along with the crop fertilizer. Rye is seeded after soybeans and edible beans, always taking advantage of a crop growing on the ground whenever possible.

As for tillage, a lot less mold board plowing is done. The chisel is used more and tillage has been reduced with most fields having a winter cover.

Fertilizer changes were made to create a balanced fertilizer. The land was in need of sulfur and traces which were never added before. For corn, a 15-5-5-18S plus traces was applied at 300 lbs/acre in the row using no extra nitrogen. Herbicides were banded and rates have been reduced by two thirds. The small grains received a balanced fertilizer fall applied and in the spring two hundred pounds of ammonium sulfate with red clover seed mixed in was bulk spread. For the soybeans and edible beans a row applied 10-9-10-13S plus traces was applied at 200 lbs/acre. Lastly, the alfalfa received 300 lbs/acre of a 9-10-10-15S plus traces as a spring applied fertilizer. Overall, the crops on this farm are performing well, soil structure is improving and the cost of production is decreasing.

As for the quality of the crops, this is harder to measure. Mineral balances have certainly changed, soil health has improved, and the crops "appear" healthy. Having fewer insect and disease problems with the crops does indicate this is true. We always say the real measure of quality comes from the animals that consume the food.

On Marvin's dairy farm, we have measured greater alfalfa mineral balance and quality. The cows are proof of the better quality. Prior to biological farming, Marvin was feeding his cows alfalfa, shelled corn, cottonseed, meat and bone meal, and soybean meal, along with lots of minerals. A 'burn and turn' type of ration, buying a lot of off the farm supplements and feeding lots of grain, protein and minerals.

The cows produced a lot of milk, but the costs and the herd health were a problem. Currently, a high forage ration is being fed. A corn and barley blend is fed at 15 lbs a cow per day and only 1.5 lbs of soybean supplement is used on cows giving over 85 lbs of milk. Mineral and protein supplementation has decreased and cow health has greatly improved.

This type of ration is a "feed what you grow" ration which ultimately reduces costs, improves herd health, and leads to increased production.

Although the crop production may not have increased significantly, the cost of production has been reduced and profitability has improved.

As for the future, Marvin's program will continue to get better as soil mineral balance and health improves. He may even consider switching to organic crop production to achieve even greater monetary returns on the crops he produces.

15

CROP PRODUCTION GUIDES FOR CORN, SOYBEANS, ALFALFA, SMALL GRAINS & POTATOES

This chapter will be on thoughts and ideas for getting the most out of your fertilizer dollar. I will outline the growth of the plant, discuss when it needs the nutrients, and cover how to put together a fertilizer and soil program for optimum return and reduced crop failure even under adverse weather conditions.

Corn

Corn is a crop grown in every state with many varied local methods. In some parts of the country zinc is included in the fertilizer program. In some areas, the use of ammonium sulfate is popular. In some places, farmers plant 30,000 plants per acre, and in other areas 25,000 is more common. Some parts of the country use no starter fertilizer, just bulk spread with anhydrous knifed in. Liquid is popular as a starter in some areas and dry fertilizer in others. I've come to believe that this is true, based on local key farmer successes or the particular sales companies in the area.

It appears that more balanced fertilizers are used in the vegetable-growing areas. There the farmers see the value of calcium, sulfur and trace elements for the vegetables and automatically follow a similar program for corn.

Then there is the national movement: narrow rows; no-till, row-till or conservation-till; and cheap fertilizer sources.

Equipment dealers are selling all kinds of add-ons to the planter: row cleaners, seed firmers, different-style press wheels, etc. Talk is of soil-seed contact and picket-fence planting. Yield is kernels per acre. You can vary the weight and size of the kernel, and the farmer's soil management and fertility program can affect these.

Nationally, we hear all about biotechnology, chemical mixes and plant genetics improvement. Look how corn is selected — it's planted in cold, hard soils; bathed in salt fertilizers; washed with herbicides and insecticides; and gassed with ammonia. It's amazing we can get any of it to grow. Do you really believe genetics and biotechnology will be able to overcome sick, out-of-balance soils? The genetics, the biotechnology and chemicals are in many cases band-aids to problems — not solutions.

What if the conditions were ideal at planting? With ideal soil temperature, soil that was loose and crumbly with lots of soil life and a balance of minerals? Now would you use the same genetics? Do you need all those extras that cover up soil problems? Some of the new technology will be really beneficial, but some will come back to haunt us, as has happened before.

You need to think of the risk. As for selecting corns from variety plots, it can give you ideas, but your farm is not their plot. If you select the corn that did the best this year, next year it may be the poorest.

On my farm, at this time I see no need for the extra special genetics or biotechnology. Our genetic selections are based on what conditions we have and the expected outcome. We use regular, popular corn genetics on the farm. We are doing research on varieties that are more workhorse types, that grow larger roots and recover more of the nutrients and moisture in the soil. We also want plants that start out rapidly and are aggressive. When feeding to animals, we look for highly digestible, high-yielding plants for silage and are presently experimenting with high-oil, waxy corns for grain. Our farm's feeding programs for dairy are based on high quality, high percent forage diets. We use low levels of grains and are trying to concentrate the energy and digestibility. I use shorter-day corn because I plant later, trying to hit ideal soil conditions. I plant my corn thicker — up to 30,000 population. Because I rotary hoe, burn and cultivate for weeds, I have to start with a few extra corn plants. I also believe I have potential for more yield with my healthy soil and balanced fertility. In many cases, low plant populations are not the farmer's limiting factor. I do believe that if everything else in the soil is "ideal," putting more seeds in the ground will result in more yield.

To start with a corn-growing program, I need to start in the fall. Managing the decay of crop residues, soil correctives and soil structure (tillage if necessary), I need to start as soon as the previous crop is harvested. No-till in many cases violates the proper management of residue decay. Residues are food for soil life and contain nutrients that need to be released for the next year's crop. Places in the Midwest are now using insecticides for corn rootworm, even when following soybeans. Without proper residue decay, insects and diseases are going to be bigger problems. Upon field inspection, volumes of corn residues from two years prior can be found.

I certainly don't think you need to bury the residues out of sight. I like to incorporate them shallowly in the soil. A rotavator, sharp disk, chopping them or running an aerator-type tool over them does wonders. If there is a high-carbon crop such as straw or corn stalks, the use of some nitrogen or manure is essential. For corn stalks, spraying on five to 10 gallons of 28 percent nitrogen with humic acid and shallowly working them in works great. You could also bulk spread ammonium sulfate at up to 200 pounds per acre in place of the liquid nitrogen.

If you are using manure, liming or making a soil correction, tillage is essential. If during the growing season, either at planting, cultivating or harvesting, conditions were too wet and soil compaction occurred, you had better subsoil. Do your evaluation.

We made our soil fertility correction, soil structure correction, managed the residue decay, and now if there is any way we can fit in a green manure crop, we do so. After soybeans and before a corn crop, we plant winter rye. When the rye gets eight to 12 inches tall in the spring, we work it in. This not only holds the soil, it feeds the soil life and holds extra soil nutrients. Row cleaners on the planter assure good soil/seed contact with all the residues.

Another place I like to seed cover crops is in growing corn. At last cultivation, we seed a blend of rye grass, red clover and hairy vetch. In some years and in some spots in the field, it does better. If the corn is poorer in an area, the interseeded crop does better; it is fixing the soil and feeding the soil life so those areas improve in the future.

As for tillage, I do believe we can do less than has been done, but here again, every farm and every soil is different. Don't copy what someone else does. Evaluate your farm for its needs.

To get good yields without ideal weather conditions and high fertilizer use, you need to grow roots. Make sure the zone you plant in is ideal for root growth. What stops or inhibits root growth is hard, tight soil, high concentrations of soluble nutrients, and chemical combinations. For the hard soils, if you don't want to work the whole field, put something on the planter and at least work the area of seed placement. Using different types of fertilizer materials that are not as soluble and are low-salt will allow roots to grow. Then by adding root stimulants such as kelp and humic acid, improvements can be made.

Corn, the Plant

Corn, commonly called maize in much of the world, is America's most valuable agricultural crop, with the United States producing nearly half the world's corn. Corn is a member of the large plant family, the grasses, to which other important crops such as wheat, oats, barley and rice also belong.

A corn plant is a marvel of energy production and storage, capturing the sun's energy during photosynthesis and converting it into food molecules. In only three or four months, a single kernel of corn grows into a plant seven to 12 feet tall and produces 600 to 1,000 kernels similar to the one that was planted.

Corn grows best in warm, sunny weather (75 to 86 degrees Fahrenheit), with well-distributed moderate rains (or irrigation with 15 or more inches of water per season) and 130 or more frost-free days. For optimum growth *and* for top quality, the growing corn plant needs an adequate and balanced supply of *all* the essential nutrients, and it needs them throughout the growing season. However, the peak time of nutrient need is in the middle of the growth cycle, when the greatest vegetative growth occurs, as well as during the reproductive and grain-filling period. The following tables give an idea of corn's nutrient requirements.

CORN NUTRIENT ACCUMULATION PATTERNS

Nutrient	Growing pattern	Leaf Translocation	Peak time of availability
Nitrogen	Throughout season, with two peaks	Yes, during grain development	Tassel to silk through grain fill
Phosphorus	Steady accumulation until maturity	Yes, during grain fill (more than N)	Early tassel and early dent

Potassium	86 percent accumulated by silking	Very little	Early tassel
Calcium	Vegetative growth (86% before blister stage)	No	Early tassel
Magnesium	Throughout growing season	Very little	Steady supply all season
Sulfur	Throughout growing season	Very little	Steady supply all season
Boron	Throughout season, with two peaks	Very little	Steady supply all season (tassel to early grain fill)
Copper	Steady accumulation throughout season	No	Steady supply all season
Iron	Two distinct times (70%) accumulation by blister stage)	Very little	Early through blister stage

TOTAL NUTRIENT ACCUMULATION
in pounds, 200 bushel per acre corn crop

Nutrient	Grain	Stover
Nitrogen	150	116
P_2O_5	87	27
K_2O	57	209
Magnesium	18	47
Calcium	4	38
Sulfur	15	18
Zinc	0.18	0.37
Manganese	0.12	1.75
Iron	0.17	1.10
Copper	0.07	0.06
Boron	0.15	0.06

Compiled from various sources

Compare the above nutrient accumulation patterns with the typical fertilization pattern for corn. Most farmers apply all the season's fertilizer at or before planting, yet the seed and the 25-day-old plant need little or none. Peak nutrient demand begins (slowly) after about 30 days and increases until about 60 or 65 days, then declines until maturation.

What happens to the farmer's inexpensive, typically highly soluble fertilizer between the time it is put in the ground until the plant is old enough to use much of it? Being soluble, some of it will leach below the root zone (if rainfall is average), some will be tied up on soil or organic matter (with little or no plant availability), and some will stay available until the plant needs it.

Fertilizing Corn

Even if you don't use all the soil-building opportunities, improvements can be made by just changing fertilizer practices. There is a lot of evidence that just getting the chemistry right will improve soil structure and biological activity. At the beginning of Chapter 13, I listed some negatives and positives. Always, do as many of the beneficial things as you can.

As you can see from crop removal and nutrient use tables, applying a highly soluble fertilizer at or before planting and assuming the fertility job is done is a mistake. We believe it is also a real mistake (not only economically, but also for health and nutrition) to follow standard crop removal recommendations by applying high levels of only three elements (N, P and K) and not giving consideration to the source of these nutrients. If you think you must put on what the crop removes, put on all elements at the level they were taken off.

Keep in mind the three main keys to using crop fertilizers:

1. Balance — provide all essential nutrients in a crop-use balance.

2. Concentration — put the nutrients in the spots where they can be easily recovered.

3. Recovery — this involves roots, large active roots with mycorrhizae living symbiotically — use a fertilizer that releases nutrients throughout the growing season, and in forms the plant can use.

Some thoughts about fertilization: Have your complete soil test done. Schedule in a soil corrective. Budget your money. First things first — that's calcium and phosphorus. Even if you can only spend $20.00 an acre, buy the material that fits your soil's needs and purchase what you can.

If soil corrections can't be made and soluble calcium has not been used on the farm, fertilizers will need to have added calcium. Pelleted lime or gypsum, depending on the situation, work well. For providing the balance of other nutrients, I will divide the soils into high-testing, intermediate-testing and low-testing soils, and I will give example fertilizer programs for each. I'm including a copy of a soil recommendation sheet. By grouping the soils in these three categories, you can monitor the whole farm and see which fields could use manure, if available. Also, the low-testing fields may be a good place to start with soil corrections. I would rather have all the fields in the medium-testing category, but balanced, than have a lot of high-testing P, K and real shortages of calcium, sulfur and traces.

If you are not sure how to divide high-testing from medium, go back and review Chapter 9. Generally, if P1 ppm is 50 or above and K is 3.2 percent or 150 to 175 ppm on medium to heavy soils, that's high-testing. Light sandy soils (CEC eight or less) never make the high-testing category; they do not have the capacity to hold enough nutrients and require more row fertilizer.

The fertilizers recommended on this sheet are Midwestern Bio-Ag's. The materials we use are: ammonium sulfate, MAP (monoammonium phosphate), potassium sulfate, sul-po-mag, Idaho phosphate, North Carolina rock phosphate, calcium sulfate, granulated calcitic lime, sea kelp, fish meal, molasses, manganese sulfate, zinc sulfate, copper sulfate, calcium borate, and humates.

I have given percents of the elements and approximate pounds per acre to apply. Note that as the soil increases in fertility, less fertilizer can be used. In the high-testing category, the amount used went down, but soluble phosphorus went up. Even if the soil is high, phosphorus is extremely important for early root development and growth. Soil phosphorus doesn't move; the root has to get to it, so the slow-release phosphorus in this mix is reduced. The soil has plenty once the root can get to it.

These soils perform well even with a liquid pop-up. Three gallons of a high quality, low-salt liquid, say 9-18-9, can get the plant started. These materials are not a fertilizer program. They don't correct soil deficiencies. They need to have the calcium, sulfur and traces added. If you used enough to build the soil fertility, cost would be prohibitive.

The reason we work with high quality dry blends is because of what you get for your money. We can balance the elements, provide enough and still compete with liquid fertilizer program costs. In our soils further north, I like adding from one to three gallons of a complete liquid plant food, with humic acid and water added for an eight-gallon minimum application placed in the seed furrow. With this same application, we apply the proper dry blend as close to and below the seed as we can, so as not to disturb the soil-seed contact when planting. This addition helps get the seed out of the ground and roots developed fast.

Nitrogen Use:

As with herbicides, I put supplemental nitrogen in the same category of "necessary evils." Use as little as possible to get the job done, always evaluating need and amount to use, the objective being to use less all the time.

Soil Recommendations

MIDWESTERN BIO-AG
10955 Blackhawk Dr. • Blue Mounds, WI 53517
(608) 437-4994 • FAX (608) 437-4441

Name:

Date:

Field Number	Soil Correction		Standard Fertilizers
	Minimum	Ideal	
High Testing P & K Soils: **Main Concerns**			**Corn** - 150 lbs. 9-18-3-9S high trace + nitrogen OR 300 lbs. 15-9-2-16S high trace + rotation & manure **Soybeans** - 150 lbs. 12-9-3-13S balanced bean fertilizer **Small Grains** - 200 lbs. 15-5-5-18S with traces **Alfalfa** - 150-200 lbs. 5-8-12-13S or 4-15-15-10S balanced alfalfa fertilizer OR custom mix calcium sulfate + trace elements
Medium Testing P & K Soils: **Main Concerns**			**Corn** - 200 lbs. 10-9-10-13S with traces + nitrogen OR 350 lbs. 15-5-5-18S with traces + rotation & manure **Soybeans** - 200 lbs. 10-9-10-13S balanced fertilizer **Small Grains** - 250 lbs. 15-5-5-18S with traces **Alfalfa** - 200-250 lbs. 5-8-12-13S with traces OR 4-15-15-10S with traces OR 3-6-20-13S with traces plus light coat of manure in fall
Low Testing P & K Soils: **Main Concerns**			**Corn** - 250 lbs. 10-9-10-13S with traces + nitrogen OR 400 lbs. 15-5-5-18S with traces + rotation & manure **Soybeans** - 200 lbs. 10-9-10-13S balanced fertilizer **Small Grains** - 300 lbs. 15-5-5-18S with traces **Alfalfa** - 250-300 lbs. 5-8-12-13S with traces OR 4-15-15-10S with traces OR 3-6-20-13S with traces plus light coat of manure in fall

When asking farmers what source of nitrogen they use, 95 percent respond by saying, 28 percent solutions, anhydrous ammonia, urea, etc. My answer would be:

1. I have some manures in my rotation.
2. I work down rye and other green manure crops.
3. I use crop rotations with nitrogen-loving crops and legumes rotated.
4. I have 25 earthworms per cubic foot of soil, and on an acre basis, that's approximately 75 units of nitrogen per acre per year.
5. My soils have 2.5 percent organic matter and are loose and well-aerated. Many more symbiotic organisms provide extra nitrogen in the soil.

So the question is, how much extra nitrogen do I need to purchase for the situation I am in now? Excess nitrogen is a negative. It slows down biological activity (fewer nodules on alfalfa and beans), makes the soil "lazy" and takes extra carbon to balance it out, meaning it could be burning up your soil organic matter. Extra nitrogen also takes calcium out of the soil. Insect pressure, diseases, plant imbalances (they slow dry-down in corn) are all affected by excess nitrogen. Other results are environmental pollution and lower feed quality. A carry-over of nitrogen from corn to next year's alfalfa is not a benefit.

You should do all the practices to grow and take advantage of free nitrogen. Nitrogen is an element you can "grow" and don't need to buy, or at least you can reduce purchased nitrogen. Not knowing for sure the kind of season you are going to have or how much you really grew which the crop could actually use this year, how do you decide what and how much nitrogen to purchase? I think it is simple and involves no figuring (it's just a guess, anyway). Nitrogen use is not a numbers game.

The Nitrogen Rules Are:

1. Split applications whenever possible if you need more than 60 purchased units.
2. Put the nitrogen where and when it is to be used and needed.
3. Use additives to reduce losses of applied nitrogen. I don't like the biological-suppressing kinds; they probably suppress other soil biological systems, and I believe the risk in using them is too high. My preferred addition is a carbon source; I prefer humic acid. My rule of thumb is one quart per 20 gallons of liquid nitrogen.
4. Trade some of your nitrogen units with sulfur; the minimum to use is a 10:1 ratio N:S in application.
5. Do add calcium to your fertilizer program. It stimulates roots and biological activity.

6. Don't plant in hard, tight soils; anaerobic organisms denitrify your supplement and send the nitrogen into the atmosphere as gas.
7. Never drive on soil when and after the nitrogen has been applied. Denitrification will take place because of soil compaction.
8. Grow large roots. What stops root growth is hard, tight soils with too many soluble nutrients and a shortage of root feeding materials.
9. Provide a balance of all the minerals. For example, trace element manganese aids in nitrogen uptake.

My preferred purchased nitrogen source is ammonium sulfate. We usually put this in the row fertilizers. For many farmers with rotations, manures and healthy soils, this is all you need. We place it where it is needed, but a little earlier than needed. Ammonium sulfate has a 1:1 nitrogen-to-sulfur ratio and is a slower-release material. This source is limited to about 50 pounds nitrogen per year, based on returns for your money and amount of sulfur wanted.

If I want more nitrogen, I like using liquid ammonium nitrate sources or URAN blends. These materials allow me not to have to violate the above rules of nitrogen. I can side dress at cultivation or if available, put them through the irrigation system, often at lower rates with humic acid added. For corn, I generally recommend 10 to 40 gallons side dress.

To side dress, you need saddle tanks on your tractor. These same tanks can be used to band herbicides (if needed) at planting; however, you will then be forced to cultivate for weed control. At the time of cultivation (foot-tall corn or so), side dress the liquid nitrogen with the humic acid or other beneficial materials added. Vary the rate, and you can determine what level works best for your farm. While doing this cultivation, inter-seed a clover/grass blend, making your trip over the field pay in many ways.

The more nitrogen production methods you have in place, the less you need to purchase. Keep checking your system for what responds for you. Many farms with medium to heavy soils, without livestock, on a corn/soybean rotation with some green manure crops, see no benefit of added nitrogen above 20 gallons per acre in addition to their crop fertilizer. Moving to lower nitrogen use and growing high-yielding, high quality crops is definitely achievable.

If you are unable to match what I call the ideal situation and side dress, there are some things and methods you can use which will help. Farmers want us to tell them how much they need to use, and they want to surface-apply it or bulk spread in early May before the corn comes up. There is no way I or anyone else can guess that amount. As with herbicides, salesmen and crop advisors in these and other situations recommend very high levels at your expense. They don't like taking responsibility for potential crop failures. You, the farmer, need to become knowledgeable and make the management decisions. Follow the practices to improve the soils. Set yourself up to be flexible and put the nitrogen on where and when it is needed. This not only reduces risk of poor crops, but puts you in charge of your own success, taking your own responsibility for success or failure.

If N is N

We are told it doesn't make any difference where the nitrogen comes from; just the amount matters. If this is so, then get your nitrogen from clovers, compost, soybeans, manures and earthworms, and if some more is needed, use ammonium sulfate. Compare these practices over a few years with using anhydrous ammonia. Maybe the yields will be the same, especially in good years, but watch the soil, its changes, and plant health. The sources may not make any difference for yield with optimum to excessive nitrogen use, but soil health will. Now start cutting back on nitrogen. The commercial nitrogen fields will suffer. They are dependent — hooked on nitrogen. With time, they will also improve if following the biological program.

With all the new technology, we are in a rapidly changing world. The global positioning and yield monitoring systems are tools that have the ability to point out spots or areas on your farm that perform better. With further search, farmers have discovered that it was *not* the phosphorus or potassium levels that gave the higher yield; it was the calcium:magnesium ratio, soil health and soil structure. These yield variations were from 80 to 200 bushel corn, and all had the same nitrogen applied. Don't be led by technology; select and use appropriate technology, and always with caution.

These are guidelines, ideals. To help formulate a biological program for your farm, you need to start from where you are now. Livestock farms grow forages, have rotations, have manure and in many ways, they are already halfway to being biological. Most use too much nitrogen and potassium and may have enough phosphorus, but not enough calcium, sulfur or trace elements. These are easy changes; so is speeding up the rotation and making better use of livestock manure. If there has been no rotation, no manure and no balanced fertilizers, and the soil is hard and dead, it will probably take longer to change or see changes. You have to slowly move from one system to another. Look, evaluate and know where you are going.

Soybeans

This is a legume crop, which means more calcium, sulfur and certain trace elements are needed. It takes four pounds of nitrogen per bushel of soybeans. For a 60-bushel crop, that's 240 pounds. You hear of farmers adding extra nitrogen and getting yield responses. If carry-over nitrogen and soil conditions are not favorable to root nodule formation, adding extra nitrogen probably does benefit, but I think that is a mistake. Correct the problem and yield responses won't occur with added nitrogen.

Soybeans like well-drained, healthy soils and don't respond well to added soluble phosphorus or potassium. We add some to the crop, along with some nitrogen. The nitrogen source we use is ammonium sulfate. It not only has a warming effect on soils, but provides some stable nitrogen to get the plant off to a good start until nodules develop.

You need to check root and nodule development during the growing season. Because the nodules with the nitrogen-fixing bacteria are so important, inoculate the beans with a quality inoculant at recommended rates.

On the crop fertilizer recommendation sheet, in the corn section you will see recommended soybean fertilizers. If you drill soybeans, a custom blend could be put together and bulk spread to address some crop needs and do a soil corrective. I like to get some kelp, root stimulants and trace elements in close to the row. Molybdenum is extremely important for nodule formation. We use Idaho phosphate in our soybean blends to provide it, as this natural material is a good source of molybdenum. Getting these materials close to the seed when drilling soybeans can be difficult. Seed treatments may benefit in this situation.

I row the soybeans because we don't use herbicides, and weed control would be difficult without being able to cultivate. Herbicides do have a suppressing effect on soybeans. They grow much taller and look different when no herbicides have been used. Soybeans do well in healthy biological soils, with balanced fertilizers. On a farm program, if money is tight and soils fairly fertile, our only fertilizer use may be 500 pounds per acre of BioCal, a soluble calcium, sulfur and boron material.

Small Grains

A recent article in a Wisconsin paper gave the results of a biological farmer who was not only providing balanced nutrients, but applying at different times in the life of the plant. His yields were up to double the area's average. Comments from farmers were shocking, "Oh, he lives near the lake, and small grains do better in cooler climates," or "It was just an accident; he couldn't repeat it," or "Look at the extra stuff he used; it must have cost money," or "He must have had a good friend running the weigh wagon."

I say he had good soil structure, good biology, proper planting and he provided the correct minerals for optimum production. Two trace elements that were used in extra amounts were zinc and manganese. Manganese is an element that will be provided by biologically active soils. Small grains don't do as well with excess soluble nutrients and are chemical-sensitive.

Farmers grew 100-bushel oats 25 years ago. Now many have given up or take the crop off as a forage. I know of three farms in the Midwest that recently produced from 100- to 125-bushel oats. Things they all had in common are that they farmed organically, which means no soluble nutrients. They have good rotations and have been doing soil corrections. Finally, they all planted heavier than normal by mistake. If soil conditions are ideal and nutrients are there, these higher populations crowd out weeds and provide opportunity for higher yields. The plants were healthy, had high test weights (35+ pounds) and didn't lodge. Lodging may be caused by an excess of nitrogen or potassium and low calcium.

At our farm in July of 1997, we had a severe wind storm which flattened the corn, but the adjoining field of barley, just two weeks from harvest, was standing perfectly. There are many things we don't know or understand. If you are having trouble with small grains, that is a clue about your soil's health and balance.

Looking at the soil recommendation sheet, in all cases we used 15-5-5-18S plus traces as the crop fertilizer. Alterations or additions could be made if necessary, but I don't like to use any other commercial nitrogen from any other source.

Planting oats on a winter-killed alfalfa crop performs wonders. Oats is a nitrogen scavenger crop. I like to plant in the fall after corn silage or sweet corn, or some early-harvested crop. This, in our area, needs to be done by Labor Day. If you have applied manure or have any soil solution fertilizer nutrients left, the oats suck them up. With excess manure and fertility, oat forage crops make poor feeds. They have such high levels of certain minerals, especially potassium, that cattle health problems follow.

The nutrients that were sucked up from the soil can't leach and are stored for future biological and crop use. Planting oats back into this dead, decaying oat crop provides conditions oats like to grow in.

A winter wheat fertility program that I have seen wonderful results with is: in the fall at seeding, select a balanced fertilizer that addresses soil shortages. This blend may need to be applied at 200 to 400 pounds per acre. In the spring, we bulk spread a 15-5-5-18S, a blend with extra nitrogen from ammonium sulfate. To this we add red clover seed. The ammonium sulfate not only has a balance of N:S, but is a stable, slower-release fertilizer. This gives us a good wheat crop, plus provides a clover cover crop for soil building.

Alfalfa and Other Forages

As a dairy nutritionist, alfalfa was the first crop I worked to improve. We had dramatic results — in feed quality changes, yield, longevity, and livestock health and performance — far beyond our expectations. The success of our dairy program comes from feeding high levels of these forages, with less grain.

Alfalfa, being a legume, likes its cations with calcium being the king. Alfalfa will grow and produce tonnage with potassium, but if you don't cut it in the bud stage, you get big hollow stems, poor fiber digestion, lower palatability and stands that don't survive.

The soil needs to be loose, crumbly and well-drained. In the harvested product, we want close to a 1:1 calcium-to-potassium ratio. This is certainly achievable. When calcium levels come in line with potassium, you don't need to cut as early, and quality will be maintained. You have a longer window of harvest, and weather at harvest will affect quality less.

I like seeding some grass with alfalfa, for better soil health and feed quality. I like using perennial rye grass. Rye grasses are more digestible than other grasses that have the same feed test. Rye grass does not overpower alfalfa and matures with alfalfa, not as other grasses do. Perennial rye grass is not supposed to be winter hardy. We do know that we can minimize this risk with proper fertility. Healthy soils with the addition of calcium and sulfur improve both the alfalfa and the grass.

Forages are grown for feed and any time we produce a forage crop, we need to supply sulfur. Sulfur is an anion and leaches, so it needs continuous additions. I want a minimum of 25 pounds of sulfur per year and want it from sulfate sulfur sources — ammonium sulfate, potassium sulfate and calcium sulfate, to name a few.

For calcium additions, lime does not supply enough. You need soluble calcium. In an earlier chapter, I outlined test results on soluble calcium sources. Farmers that grow forages usually test them for nitrogen (from which protein is calculated), fiber (from which energy is calculated) and minerals. This is a wonderful guide to your fertilizer and manure management.

Nutrient	Soil notes	Alfalfa content Avg. conventional level*	Alfalfa content MBA reasonable level	Difference
Calcium	Normally ignored; consider only pH	1.2%-163 g/day**	1.8%-245 g/day**	+82 g
Phosphorus	Requires "healthy" soil for uptake	0.26%-36 g/day	0.38%-52 g/day	+16 g
Magnesium	Often applied in excess, causing tie-up, hard soil	0.32%-43 g/day	0.50%- 68 g/day	+25 g
Potassium	Usually applied in excess	Normally excess with no yield advantage when over 2%	Targeting 1:1 Ca:K for quality	
Sulfur	Often ignored	0.23%-31 g/day	0.36%-49 g/day	+18 g
Trace elements	Often ignored	Not part of conventional fertility program	Home-grown are bio-available (organic/chelates)	

142 g total

* Many biological farms maintain these levels

** Grams per day based on equal dry matter intakes of 30 pounds of either forage, assuming a 60 percent forage ration for a cow consuming a total of 50 pounds of dry matter.

The report below is from Mike Lovlien, a Midwestern Bio-Ag consultant from Minnesota. Mike compared 16,000 feed tests performed by Dairyland Labs, from farmers all over the Midwest, to 295 samples from farmers using our balanced fertilizer program.

Fertility Programs and Mineral Uptake of Alfalfa by Mike Lovlien

Results from a compilation of data from an independent laboratory concluded that on average, haylage samples taken from fields managed under Midwestern Bio-Ag's program contained 47 percent more calcium, 16 percent more phosphorus, 32 percent more magnesium and 29 percent more sulfur than the average of all other (non-MBA) samples from that same laboratory.

Test	Aver. Dairyland Labs mixed haylage samples		Midwestern Bio-Ag farm samples using MBA fertility program		
	1995	1996	1996-97	Change	MBA desired
No. samples	16,662	19,984	295		
Crude protein %	19.39	19.57	20.50		18 - 21
Insoluble protein %	1.82	1.81	1.62		
Protein solubility %	49.38	49.27	50.63		
A.D.F. %	35.43	34.95	33.39		28 - 30
N.D.F. %	45.13	44.52	43.27		A.D.F. + 15
Calcium %	1.10	1.15	1.62	+ 47%	over 1.5
Phosphorus %	0.31	0.33	0.36	+ 16%	over 0.35
Magnesium %	0.28	0.29	0.37	+ 32%	over 0.35
Potassium %	2.57	2.70	2.86	+ 11%	1.5 - 2.0
Sulfur % (no. samp.)	0.24 (292)	0.28 (662)	0.31 (279)		+ 29%
N:S ratio	13:1	11:1	10.5:1		10:1
Nitrogen %	3.1	3.13	3.28		

Calcium is the "trucker" of all minerals, and we note a dramatic increase in mineral uptake when calcium levels are good. These are just numbers; we are also convinced the cows like it better. As one farmer commented, "It just feeds better." Besides palatability, the digestibility seems to have gone way up, and that is based on cow performance.

If you look at crop removal charts, you will notice that it takes more boron to grow corn than alfalfa. Then why is everyone adding boron to alfalfa and not to corn? Boron has two main functions: 1) energy translocation and 2) calcium uptake. If calcium is the trucker of minerals, boron is the steering wheel. Boron, like nitrogen and sulfur, is an anion and needs to be continually in the fertilizer program.

A minimum of 25 pounds per year of soluble calcium should be applied for alfalfa production. Calcium nitrate supplies soluble calcium at one pound per gallon. It would be way too expensive to provide all the soluble calcium this way. Lime provides two pounds per ton; again it is impossible to provide enough with this source. Under very high biologically active soil with high calcium levels, minimum levels of calcium are provided. Gypsum (calcium sulfate) supplies 25 pounds per ton, and BioCal (a hydrated lime) supplies 150 pounds per ton. These are not liming materials (to raise pH), but are just to supply extra nutrients (more like a fertilizer), above and beyond the soil's ability. Sulfate sulfur at 25 pounds sulfur per year and boron at one pound actual per year are essential for crop performance.

If P and K levels are high in the soil, we add our calcium, sulfur and boron and supply a mix of other trace elements as needed.

Getting higher levels of magnesium into the plant also improves cattle performance. Having high potassium levels suppresses magnesium uptake. Cattle symptoms are known as grass tetany. Even if your soils are high in magnesium, you may not have enough in plant-useable form. We want to balance the soil and provide a balanced diet of all essential minerals to the growing crop. Because the soil was testing high in magnesium, we ignored the need to supply additional magnesium. Adding five or six pounds per year per acre does make a difference. This is true for alfalfa and many other crops. The main source we use is K-mag or Sul-Po-Mag, a natural mined material.

Alfalfa loves cations, the positively charged elements: calcium, magnesium, potassium and sodium. What about adding sodium? I have been doing test plot work using sea salts and kelp. We believe this mix is increasing the plant sugar levels. Every element is important. We talk about the negative effects of chloride because of the excess that farmers use, but small amounts of chlorine are needed. Livestock manures provide some sodium and chloride. If you grow hay commercially with no livestock, you may need to include these in your fertilizer program.

I have been writing about fertilizer; how about management? Seed bed preparation is critical. These are small seeds. They cannot be buried deeply and expected to grow. A good firm seed bed with good phosphorus levels and abundant available calcium are necessary to establish alfalfa. To keep it going, cutting schedules and feeding are important. Because you are removing the whole plant top, large volumes of minerals are removed from the land. For livestock farmers, mineralizing and taking care of this crop are the most important things you can do to improve farm profitability.

To establish alfalfa, I like seeding with a companion forage crop. We use triticale and peas. This gives us a quality feed and tonnage. Following this crop, we must get at least one cutting of hay. I like to let the new seeding blossom so root development can be maximized. I also like the crop to be at least a foot tall going into winter. After the last cutting and before winter growth, a balanced fertilizer is applied along with 500 to 1,000 pounds per acre of BioCal, our soluble calcium, sulfur and boron liming material. If the field's soil fertility is medium to low, I would also apply a light coat of compost or manure. A slinger type of manure spreader is surely a wonderful invention. Our spreader holds four tons, and we apply between one-half and one load per acre. If dollars are short and you like to spread costs out, do the balanced crop fertilizer either early next spring or after the first cutting. The crop recommendation sheet gives you some ideas for the crop blend.

If the fertility is balanced and you are feeding the crop a balanced diet, cutting each time in the bud stage is not necessary to achieve quality feed. I like getting the first crop off fairly early. Then sometime during the growing season, I like to let the crop blossom. When the plant switches from vegetative growth to flowering, lateral roots grow.

For our own farm, three crops are all I care to harvest. We get excellent quality forages with this balanced fertilizer and do not need to make four cuttings. Our tonnage is there and so are palatability and digestibility. Making just three cuttings also allows us to go into the winter with some growth on the plants.

How do you decide when to cut? I watch three things. 1) The stems. If they are coarse and hollow, it is time to cut. Even if there are no blossoms, a coarse, hollow stem will only get coarser and hollower, with more undigestible fiber. 2) New growth coming off the crown. If the plant is sending up new plants, cut. You don't want to wait until this new growth is so large that it gets damaged in harvest. 3) The soluble carbohydrate level (refractometer reading; see Chapter 8). It is an easy field test. The refractometer measures quality indirectly. We like readings over 12 Brix (percent soluble solids), but if weather conditions are less than ideal, this may not be possible. Measuring the sugars will also prove to you the necessity of cutting in the afternoon or after several hours of sunshine.

You probably think I forgot the main reason for cutting or not — weather. Watching the weather forecast is a guessing game and not sound advice to follow. But I keep on watching; I need someone to blame when my hay gets rained on! A major benefit of balanced, biologically active soil with a lot of available calcium is that you have a longer period of harvest time before crop quality begins to deteriorate.

Alfalfa chokes out from hard, tight soils or starves out from poor fertility practices. If you constantly remove all the organic matter, what is left to feed the soil life? That is why (besides adding nutrients) we apply a light coat of manure or compost in the fall: we are feeding the soil life.

As for hard, tight soils, we do drive on the field many times harvesting. Even if we have good, biologically active soils, we sometimes harvest when the soil is not as dry as it should be, and compaction can occur. That is one reason I like to add grasses to my forage mix. I have had success using an Aerway aerator to loosen soils. When that is done, the soil has many holes poked in it. This is also beneficial to do before adding the manure or compost, to reduce runoff. We call it our mechanical earthworm. If your soils have an abundant supply of earthworms, this practice may not be necessary.

Potatoes

Why potatoes? Being a dairy nutritionist, how did I ever get involved with potatoes? About five years ago a potato grower in central Wisconsin rented a field from a biological farmer that had been on our program. The crop he grew got his attention, and he wanted to know what we were doing.

It doesn't matter what crop you grow; the principles are the same. They all need balanced fertility; and some crops have specific needs. To learn about potatoes and farming irrigated sands, I hit the books and the road. I checked out the university recommendations and then went to the farms and consultants and found out their practices. Then I checked with growers who were on the fringes, doing unconventional things; some working, some not. I also traveled to Idaho. If you want to grow potatoes, go where they have a reputation for good potatoes. Idaho potatoes are grown on totally different soils than here in Wisconsin. They have high calcium, high pHs, large CECs and hard, tight soils. The climate and natural soil minerals may be giving them their reputation. While in Idaho, I visited Stukenholtz Labs, a soil and plant tissue testing lab and consulting firm. Dr. Dale Stukenholtz gave a wonderful overview of Idaho potato growing and had tests and research to support his work.

The usual practice in Idaho is humic acid use. This material is an industry standard. It is added to nitrogen and liquid fertilizer mixes. The soils are hard, and the farmers believe the humic acid loosens their soils. Many of them also V-rip in both directions before planting. Besides providing balanced nutrients, they grow in what I would call a bath of nitrogen — 300 to 350 pounds per acre. This seems excessive, but maybe those tight soils had tremendous losses. Without soil air, denitrification takes place rapidly. It doesn't take 300 pounds of nitrogen to grow potatoes, but that is what they use to get the job done.

Another interesting point was on calcium. Even with their high-calcium soils, the amount of plant-useable calcium was low. This may be partly due to the high nitrogen use, but also just because it shows up on a soil test does not mean crops can use it. For the hard, tight soils, I suggested green manure crops, but due to limited water and irrigation costs, it seemed prohibitive.

Back to Wisconsin. In this state we plant different varieties of potatoes than in Idaho. We also have very different soils. Most potatoes in

Wisconsin are grown in the Central Sands irrigated area. These light soils, with CECs of eight or less, seem to be the medium in which to grow these hydroponic plants. Many farmers here do not and will not add lime. The soil pHs are 5.0 or less. They believe this low pH reduces scab disease on their potatoes. Telling this story in Idaho gets a laugh. Their pHs are 8+, and they believe scab is caused by decaying organic matter lying too close to the potato. They grow different varieties, and this makes a difference.

So what's the truth? What does seem obvious here in Wisconsin are the many benefits of adding calcium. The field that was rented from the biological farmer and grown in potatoes had a pH of 6.8, with a 75 percent base saturation of calcium.

Since 1985, research in Wisconsin has shown many benefits from supplying soluble calcium. I am including a summary of some of this work as reported in the *Badger Commentator* magazine.

"Calcium Contributes to Healthy Harvest: Later Season Calcium Additions Improve Potato Quality And Protect Yield In Hot Weather" (*Badger Commentator*, June 1997, pages 12, 14).

> In the mid-1980s, University of Wisconsin-Madison horticulturist Jiwan Palta began urging potato farmers to add water-soluble nitrogen and calcium to their fields in August. His advice was greeted with much skepticism.
>
> "They thought I was crazy," Palta says. "Other professors here in the college questioned the strategy. At that time, the college recommended that farmers add calcium if soil tests showed there was less than 800 pounds of calcium per acre. But our research found improvements in potato quality when we applied soluble calcium to fields with more than 1,500 pounds of calcium per acre."
>
> Now, Palta has earned substantially more respect. Colleagues invite him to present papers at national symposia. Scientists in Idaho, the nation's potato capitol, want to collaborate with him on research. And today, a lot of Wisconsin potato farmers follow Palta's advice. The reason is simple. Mid- to late-season fertilization with calcium improves their potato yields, quality and their profits.
>
> In the early 1980s, UW-Madison plant pathologist Arthur Kelman and others found that tubers with more calcium were less susceptible to soft rot in storage. Then Palta, a plant physiologist, discovered that potato tubers have small roots. That finding, which Palta and his research assistant published more than a decade ago, changed the way scientists think about the mineral nutrition of potato plants.
>
> "You have to begin thinking about the tuber as a separate underground plant," he says. "When you apply calcium to the plant's main roots, it doesn't increase the calcium level in the tubers. Growers fertilize the plant in the spring when it starts growing. When the tubers are growing in July, August

and September — what growers call bulking — the tubers need fertilizer too."

Each summer day a potato plant loses water vapor from its leaves and draws water in through its roots, Palta explains. With water comes calcium. But a tuber loses little water and takes up little water or calcium.

"A tuber is a modified underground stem," Palta says. "But the concentration of calcium in the above-ground plant stem is five times greater than that in a typical tuber — about 100 parts per million."

During the past decade, Palta showed that the more calcium he can get into potato tubers — until about 300 parts per million — the healthier they are.

Palta measures potato health by improvements in storage life and tuber quality. Farmers can lose from four to 25 percent of their crop in storage to bacterial soft rot. Also, the processors who turn spuds into potato chips and French fries check potatoes for internal defects, such as hollow heart and brown spot. Processors reject potatoes by the truckload if they have too many defects.

Following Palta's advice, growers now apply a water soluble form of calcium and nitrogen through their irrigation systems three times between hilling and harvest. Palta calls it spoon-feeding tubers during bulking. On average, the applications net growers about $70.00 per acre.

Palta and his research group now have found that these calcium applications also benefit the crop during summer hot spells.

The studies began in 1988, the warmest growing season in recent memory. The temperature reached 90 degrees or above on 46 days that year in central Wisconsin. Although area growers irrigated their potato fields, yields were down 25 percent across the region — an average loss of about $500.00 per acre.

While studying late-season fertilization in 1988, Palta found that the plots to which he had added calcium and nitrogen during July, August and September produced normal yields.

Native to Central and South America, potatoes grow best in moist cool climates, such as those of the Andes. "The ideal temperatures for growth are about 70 during the day and 60 at night," Palta says.

Palta tried to repeat the experiment in subsequent years, but none of the growing seasons was as hot as 1988.

To recheck the field results, Palta turned to the UW-Madison Biotron, where he could control the environment. There, Palta and graduate assistants Ahmed Tawfik and Matthew Kleinhenz grew potatoes under temperatures that peaked at 86 degrees for six hours a day over several weeks. The soil, brought from the Hancock Agricultural Research Station, contained adequate calcium for potato growth under normal conditions. The researchers added nitrogen to some plants before the heat stress, and added nitrogen, or nitrogen and calcium, to others during heat stress. The plants that received nitrogen before or during heat stress developed classic symp-

toms of calcium deficiency and yielded 1.35 and 1.54 pounds of potatoes per plant. The plants that received both calcium and nitrogen during heat stress showed no signs of nutrient deficiency and produced 2.20 pounds of potatoes per plant.

Palta can't yet explain why potatoes with added calcium do so much better. But the results suggest that plant leaves need more calcium to function properly at high temperatures.

Palta and research assistant Chris Gunter now have begun to see if increasing the calcium level in seed potatoes gets the crop off to a stronger start after planting. The initial year's results look promising, he says.

In this article, they are talking about late-season soluble calcium. With all the potassium chloride and nitrogen use and the low pH, light soils, it is no wonder that no calcium is left when it is needed. We started a project of adding our BioCal to 1,200 acres of potatoes, spring-applied at 1,000 pounds per acre. This should supply 75 pounds of soluble calcium plus some sulfur and boron. We will look at yields, but also quality. I have asked the farmers to reduce their nitrogen applications if at all possible. It would take 75 gallons of calcium nitrate to provide that much soluble calcium. If we don't get the results we expect, we will have to come up with a way to apply BioCal later in the season.

Let me outline a typical Wisconsin potato fertilizer program, and you can see if you can find limiting factors or ways to improve their program.

- 300 to 500 pounds 0-0-60, bulk spread.
- 400 to 600 pounds starter, a 5-21-25 + traces is typical. This is probably made with DAP and 0-0-60 (KCl).
- First hill: 300 to 500 pounds ammonium nitrate/ammonium sulfate blend.
- Second hill or with irrigation: extra nitrogen, up to 200 units total for the season.
- Early July or end of June: some calcium source, either calcium nitrate or gypsum, or both.
- Some farmers will foliar feed with a plant food blend during the season.
- In addition to their fertilizer program, many farmers were fumigating the soils before starting, spraying a fungicide weekly, and using insecticides and herbicides on a regular basis.

I was shocked. I am used to working with corn and soybean farmers who want to keep expenses down to $30.00 an acre for fertilizers. My first observation was, we sure have room and resources to work with. I also looked at potato crops that didn't follow this program: more biologicals used, reduced fertilizer rates, ones that limed, reduced nitrogen rates — and some of these growers were having successes.

Here are some of my suggestions for the potato grower in the Midwest:

1. Start with a good base of calcium. Add a soluble calcium source for season's use.

2. Reduce nitrogen rates to 125 to 150 units. This seems to be an adequate amount if split applications are used, and if we add humic acid to the liquid nitrogen to reduce losses. Most farmers had been using some ammonium sulfate in their programs, which I encourage.

3. These are light sandy soils, and magnesium additions seemed essential. The petiole test on one of the highest-yielding potato crops in Michigan had high magnesium and high boron levels, both of which are essential for top yields and quality.

4. I suggest a potassium blend, instead of just 0-0-60. I had them blend a mixture of 0-0-50-17S (potassium sulfate), K-Mag (potassium-magnesium sulfate) and 0-0-60 (potassium chloride). There is a lot of research showing high specific gravity and tuber dry matter increases with potassium sulfate over potassium chloride. Potassium sulfate is also slower release, giving more season use, and because I believe calcium can replace some of the potassium used, we reduced our application rates.

5. Some trace elements had been added, but at low rates, and just a generic balance. I increased the trace element fertilizer, used a homogenized blend, and added extra copper and boron. Boron is essential for sugar translocation.

6. I suggested adding humates and other carbon sources to their starter fertilizers to provide absorption sites to reduce leaching. We then reduced the starter fertilizer to 400 pounds per acre.

7. Instead of all soluble nutrients in a sandy soil, I wanted to add some slow-release natural materials. K-Mag and Idaho phosphate were included in their row fertilizer blend. I switched from using DAP to MAP.

8. Finally, I wanted to add some kelp (seaweed). A lot of research has been done on root development and the benefits of cytokinins for potato growth. We added kelp in the starter blend at 25 to 40 pounds per acre.

Did this add cost to the farmer's present program? In some cases we were just trading dollars, and in others the cost did go up, but the opportunity is unbelievable. I would like to be able to eliminate the fumigant ($200.00 per acre). There is some research showing that adding compost and promoting the "good bugs" may eliminate the need for fumigants. As with the fungicides and insecticides, with proper IPM (integrated pest management), these should be able to be reduced if we get healthy plants as I expect.

On an organic potato farm, this year (1997) leafhoppers were a problem. We were testing BioCal application rates and noted a remarkable change. At 450 pounds per acre, infestation was severe; at 750 pounds, insect numbers were less; and at 1,000 pounds, hardly any could be found. This was a split-field, side-by-side comparison. We can't explain why this happened, but are encouraged by these results. It confirms our belief that healthy crops are not attacked by insects.

16

FINE TUNING: BIOLOGICALS, GROWTH REGULATORS, FOLIARS

When everything is working right, nature's systems automatically cause plants to grow well, to reproduce and to remain healthy (to naturally resist pests and diseases). The soil and the beneficial organisms in it will automatically break down organic matter and soil minerals and release a balanced supply of nutrients, and they will automatically condition the soil and give it a loose, granulated structure that allows movement of air and water. The farmer would only have to put the seed in the ground, stand back and watch nature's systems produce an abundant harvest.

But, usually it doesn't work that way. In the real world, completely ideal conditions seldom happen. Maybe the weather is cool and wet . . . or hot and dry. Maybe you don't have time to spread manure in that back field. It develops a crust from low humus, and then a good crop of weeds springs up and you can hardly see the crop you planted.

No matter how great the principles of biological farming sound, the realities of making them work — day in and day out — are not always easy. Any system of agriculture, whether chemical-intensive conventional farming, middle-of-the-road biological, or toxic-free organic, *all* have unexpected problems and glitches that pop up to challenge you. The successful farmer will meet them head on, devise a way to overcome them (based on knowledge of the soil-plant system, or perhaps on the advice of a consultant), and then he will do what needs to be done. It takes guts and a bit of faith in the system of agriculture you are following.

Unfortunately, some people don't really see the "big picture." They don't see what they are working toward — their goals. They may not understand how the various parts of the system fit together; for example, how different tillage methods affect organic matter decay, soil microorganisms and soil structure. If you don't understand how the soil-plant system works, you may wind up frantically grasping at straws — trying this or that new method or "magic in a gallon jug" some salesman talks you into.

By understanding the whole system, you are better able to diagnose the real causes of a problem and then devise a solution that attacks the causes, rather than slapping on "band-aids" that only treat the symptoms of the problem. For example, many people think an aphid infestation is the problem that is reducing their alfalfa growth. They roll out the spray rig, fill it with super-toxic "nerve poison" and blast those aphids into oblivion — along with the ladybugs and green lacewings that were busy eating the aphids. This "quick-fix" approach to farming doesn't trace the pest attack back to sick alfalfa plants (unable to produce aphid-repelling substances), because of tight, compacted soil (restricts soil aeration, causing sick roots and sick plants), that resulted from the use of high rates of humus-depleting high-salt fertilizers over the last 10 years. The real cause of the aphid infestation was poor farming practices and harsh fertilizers. The answer: *Fix the soil first.* That is the only real solution.

One difficulty is that the biological methods that ultimately will fix the problems (and prevent new ones) do take some time — generally three to five years to get an abused soil turned around. In the meantime, you have to grow decent crops to stay in business. Even after you have the soil working pretty well, a really bad spell of weather may throw a monkey wrench into the system, and you may need an emergency rescue to give a lagging crop a boost — or to "prime the pump" and get better microbial activity in the soil.

There are many products and materials on the market that are natural and non-toxic, and which may help you through a tight spot. As a general, inclusive term, I like to call them "biologicals." Some people not considering the big picture may regard them as *the answer for successful agriculture* (the "silver bullet" syndrome). I would rather view them as useful tools and fine-tuning products to help you move toward a self-sustaining biological system. Once the soil-plant system gets working (see Chapter 3), nature's processes should automatically keep giving you healthy, high-yielding, high quality crops (and healthy, productive animals). You should only need to recycle organic matter and supplement low levels of nutrients. Purchased inputs can be very low, thus greatly improving profitability. Even if terrible weather knocks growth back some, the crops grown on biologically managed soil will be a lot better than crops grown on abused, out-of-balance soil. Those are times when biologicals can also be useful, as a "booster shot," rather than a permanent part of your farming system.

Most biological products and materials have a bad reputation among conventionally oriented scientists and company representatives. Perhaps partly because they may not realize the importance of the soil's organic matter and beneficial organisms. Another reason is that most such products, when tested by conventionally oriented institutions or companies, just don't seem to perform the way they are promoted to perform. Possibly the main cause is that biological products often show little or no crop response (not "statistically significant") in good soil. Good, biologically active, well-aerated soil doesn't need the effects of the product. Nature's processes are already doing the product's job. That is why I look on biological products as occasional helpers, not full-time components. Much scientific testing is done on already fertile soil, so there will be little crop response to the biological product. If the testing was done by reducing the fertilizers in half, maybe then we would see a difference. Maybe the product was tested while still using harsh salt fertilizers which hinder the product's performance. Another reason is that the researchers usually only measure yields and do not consider crop health or quality. A biological product may not give any better yield than a strong dose of commercial N-P-K fertilizer.

With those insights in mind, let's survey the types of commonly used biological products and see what they do and how they can fit into a biological farming system. Many of these products and materials have multiple effects, both direct and indirect. They act on many parts of the plant-soil system. Perhaps a product mainly stimulates certain soil microbes. By increasing the population of microbes, additional nutrients may become available. Root growth may be stimulated (from extra nutrients plus microbially produced growth stimulators). Thus, top growth and total yield and quality may be raised, plus soil structure may be improved (decreasing soil erosion). That's quite a performance for some murky-looking liquid in a jug. Many times you are unable to measure or actually see the improvements, if there are any. You can see why the scientists are often skeptical. Again, if you understand the workings of the soil-plant system, you can see how many good effects can come from improving just one part of the system. You can realize that if the system is already working well, tinkering with it won't do much more.

Enzymes

Enzymes are called biological catalysts. A catalyst is a substance that speeds up a chemical reaction. There are inorganic catalysts, like in your car's catalytic converter (which breaks down air pollutants), and there are the organic catalysts — enzymes. Enzymes are special proteins, produced by all living cells.

Almost all life activities — from photosynthesis to digestion to growth and reproduction — operate by a complex series of chemical (or biochemical) reactions. A fire is an example of a chemical reaction. Some kind of fuel is combined with oxygen (oxidized), and heat and carbon dioxide are released. When you digest your food and it is used by your cells, the cells "burn" it, but not in the violent, uncontrolled reaction of a fire. The life processes operate in a very controlled, orderly way (otherwise chaos would result and life would not be possible). A fire requires a certain high temperature to begin burning. Living cells operate at much cooler temperatures. The chemical reactions necessary for life would not happen fast enough if it were not for catalysts which speed them up (or allow them to occur at a low temperature).

Each main type of activity — photosynthesis or cellular energy release (respiration) for example — happens through a long series of reactions, dozens of them, that have to occur in a certain order. In cellular respiration, a sugar molecule containing six carbon atoms, six oxygen atoms and 12 hydrogen atoms, is slowly "whittled away" (broken down) a little at a time until it is changed into water and carbon dioxide. Along the way, some of the energy stored in the sugar's chemical bonds is released, and that energy is what "runs" all living cells. At almost every step (reaction) along the way, a certain enzyme is required to make the reaction occur at a reasonable rate.

There can be one or two thousand different kinds of enzymes in each microscopic cell of a plant's or animal's body. Each kind of enzyme only catalyzes one specific type of reaction. For example, one enzyme is required to link a glucose sugar molecule to a fructose molecule and form a sucrose molecule. A different enzyme is necessary to split the sucrose into glucose and fructose.

Some enzymes are composed only of a protein molecule, but many others require an additional component in order to work. That necessary "activator" is either a nutrient ion (such as calcium, magnesium, iron, cobalt, copper, etc.) or a certain vitamin (some of the B vitamins, for example). This is one reason vitamins and minerals are so important for health.

Enzymes are manufactured inside living cells, but they can also be released outside the cell into the environment. That happens when your pancreas and small intestine secrete digestive enzymes. It also happens in the soil when microorganisms (bacteria, actinomycetes and fungi) release enzymes onto the organic matter they are decomposing (digesting). They might release a cellulase enzyme, which breaks down cellulose (enzymes are usually named for the action they perform or the material they act upon). Similarly, there might be a protease that digests protein and a lipase that breaks down fats (lipids). A single bacterial cell might release 30 or more different enzymes as it "has dinner" and decomposes a fragment of alfalfa leaf.

There are various agricultural products that contain enzymes (usually free enzymes without live microbes). The products may also contain additional materials, such as a nitrogen source, trace elements or a sugar. A particular product may be intended for a certain use. Some are to be sprayed on or injected into the soil as a general stimulant of soil microbial activity (sugar also does this by acting as a food source). They would mainly act by digesting and releasing additional nutrients that were tied up in organic matter. They basically "burn up" organic matter. This will stimulate both microbes and plant growth, but it will use up the soil's organic matter unless more is recycled. Some nice yield increases have been cited from using these products, from 5 to 25 percent increases for row crops, small grains and vegetables.

Some such products are designed to especially digest crop residues and stubble, which can be slow to decompose if there is little available nitrogen to balance their high cellulose content. These products would need warm, moist conditions to do their job, and they can work well.

Any increase in soil microbial activity will generally also improve soil structure, so the claims by some companies of their product's reducing compaction or a hardpan can be true. Other benefits can be increasing the soil's cation exchange capacity (CEC) (if humus is produced) and detoxifying soil (microbes can break down most toxic chemical pollutants).

Some other products contain enzymes that can be used by growing plants. They may be applied as foliar sprays, sprayed on potatoes and fruit to increase storage life, used as seed treats to improve seedling establishment, or used as dips to reduce transplant shock. Foliar-applied enzymes are said to enter the plant, be transported throughout it and begin operating in less than an hour.

Plant Hormones and Growth Regulators

Plants grow, mature and reproduce, and these activities require certain "chemical signals" or regulators called hormones. We are more familiar with human and animal hormones, which regulate growth and several other body functions. Plant hormones are quite different. Plants manufacture hormones in their tissues in very small amounts. The hormones travel to certain target tissues and regulate certain functions. For example, hormones called auxins are released from stem and root tips, buds and flowers. Auxins control the growth of these new tissues (cell elongation) as well as the bending of a stem toward light, fruit ripening and leaf drop.

Other hormones, called gibberellins (named for one of them, gibberellic acid), regulate cell division and growth, root elongation, bud formation, breaking of winter dormancy, seed germination, leaf expansion, and flower and fruit formation.

Hormones called cytokinins regulate cell division and growth, root formation, breaking seed dormancy, bud formation and growth, phosphorus absorption and transport, flower and fruit formation, and delay of aging.

A hormone called abscissic acid, promotes bud and seed dormancy, and leaf drop. It causes the slow-down of growth in a maturing plant.

Other substances which are technically not hormones still have somewhat similar growth-regulating effects. These include:

Vitamins

Many of the B vitamins act as hormones in plants, even though they have completely different functions in humans and animals. They help regulate root growth, for example. The B vitamins thiamine, niacin, pyridoxine, riboflavin, biotin and aminobenzoic acid are especially important. Vitamins are manufactured both by plants and by certain soil microorganisms, so biologically active soil often stimulates crop growth in this way. Some vitamins also improve plant growth by being enzyme activators (see above, "enzymes").

Ethylene

Ethylene is a gas produced in plant tissues which can travel through the air and affect nearby plants. It can trigger seed germination and promote root growth, flower formation, fruit ripening (it is used commercially to ripen early-picked fruit and vegetables) and leaf drop.

Ethylene is also produced by some soil microorganisms (those that live in anaerobic, or low-oxygen parts of the soil) and is present in small amounts in most soils. It appears to regulate the activity of certain microbes, since some aerobic species are inhibited by it and others are stimulated. When raw organic matter is added to the soil, ethylene production increases. The presence of ethylene in soil appears to inhibit the germination of some fungal disease spores, including Phytophthora, Fusarium and other root rots. That is one reason adding compost or other organic matter often controls crop diseases. Being a plant hormone, ethylene from the soil can stimulate plant growth. This interesting subject was discussed by plant pathologist Dr. R. James Cook, Washington State University, in the journal *Compost Science*, vol. 17, no. 2, pp. 23-25 (reprinted in the book *Organic Farming; Yesterday's and Tomorrow's Agriculture*, 1977, pp. 157-164). Dr. Cook also notes that nitrate fertilizers and too much tillage can reduce helpful ethylene production, while ammonium fertilizers do not.

Humic acids

Humic acids, possibly along with other substances found in humus, and products derived from plant matter can have hormone-like effects on growing plants. Because of their complex and poorly known structure, it is only assumed that humic acids may act like auxin hormones. Humic and fulvic acids will be covered in a later section of this chapter in more detail. They have been found to increase cellular respiration, root growth, stem growth and crop yield.

Some application of plant hormones has been made in commercial horticulture, where gibberellins are used as root dips or seed treats for example. In field crop production, little has been done with pure hormones, but some products such as kelp and humic acid, do contain hormones or have hormone-like effects. These two products will be covered below.

Microbial Inoculants

We have seen the importance of beneficial microorganisms in the soil (see Chapter 5). They recycle organic matter, release tied-up nutrients, fix nitrogen, improve soil structure, "feed" roots, and protect plants from diseases. Quite a few products exist or are being developed that supply living microbes, presumably to supplement those already present or to supply those that should be there in good biologically active soil. There are also products designed for inoculating manure storage facilities, compost, stubble or other crop residues, and pollutant spills (such as oil spills or PCBs).

Other inoculants are for foliar application, sometimes used to reduce disease or pest attack, or to prevent frost damage.

Several kinds of microorganisms could be involved in the various products: bacteria, actinomycetes, fungi and algae. Some products contain only one or two species, while others contain many. Research has shown that microbial inoculants sometimes fail because the inoculated species cannot easily become established due to competition from other species already established in that soil. The various species of soil microbes form a "community" or ecosystem, with "give and take" interactions among them. They become attuned to each other and occupy all the available microhabitats, so an outside species has difficulty finding a niche. Many of the modern microbial inoculants, however, contain species or strains that do have good competitive abilities, so their success rate is higher than earlier products. Most products also accompany their microbes with aids to establishment, such as sugars and nitrogen (food), enzymes, trace minerals, humic acids or organic acids. Listed below are *some* of the main kinds of organisms.

Organic Matter Decomposers

Included here are bacteria, actinomycetes and fungi that normally break down raw organic matter into humus and available plant nutrients. Most soils should have sufficient populations and variety of species of decomposers so that whenever more raw organic matter is added, they quickly multiply and go to work. However, abused soils, soils very low in organic matter and soils polluted by toxic agricultural or industrial chemicals may have too low, or too few, populations of the necessary species. Inoculating with a product containing the right balance of species for the material to be decomposed will often give good results in these soils. Treating polluted soils and oil spills with microbes (a practice called "bioremediation") has proved to be a cost-effective alternative to land-filling or incineration of such materials.

Using decomposing microbes as compost starters, in manure-treatment facilities and in septic tanks and municipal water treatment facilities is a long-established practice.

Root-Inhabiting Microbes

Certain strains of bacteria, actinomycetes and fungi live mainly on the surface of plant roots; they are called rhizosphere species. They may perform various beneficial functions that aid plant growth, including increasing nutrient availability, channeling water and nutrients into roots, releasing plant growth-stimulating substances, fixing nitrogen, and protecting roots from pathogens (diseases). Some species perform more than one function. Some kinds of microbes used in products or being investigated by researchers include the bacteria *Azotobacter, Bacillus, Psuedomonas, Agrobacterium* and *Pseudomonas*; the actinomycetes *Streptomyces* and *Nocardia* and the fungi *Aspergillus, Penicillium, Trichoderma, Chaetomium, Corticum* and certain mycorrhizal species. The nitrogen-fixing bacteria in legume root nodules (*Rhizobium*) have been used for decades, and a tropical bacteria (*Azospirillum*) is being studied because it lives on grass roots. Imagine planting nitrogen-fixing corn.

Some of these root-inhabiting microbial products can be sprayed or injected into the soil, but many give more cost effective results when used as seed treats or dips (in horticulture).

Mineral Releasers

There has been research, especially in Russia, on certain soil microbes that may not always live on roots, but which are often effective releasers of the nutrients tied up in soil minerals. The main interest has been in the "phosphorus-solubilizing" species, including *Azotobacter, Bacillus* and *Pseudomonas*. Early Russian studies reported hefty yield increases, but American research has seldom duplicated those results. There has not been much development of commercial products along this line.

Foliar Species

Certain bacteria or fungi that live on plant leaves can neutralize the attack of pathogenic species. Other leaf bacteria have been found to reduce frost damage by changing the way ice crystals form. These are still developing research areas, so commercial products may be coming down the pipe.

One species of bacteria, *Bacillus thuringiensis*, causes a fatal disease when ingested by certain insects, mainly caterpillars. Foliar sprays of this microbe have been used for years to combat crop pests, and recently new strains have been developed that control beetles such as the Colorado potato beetle. Commercial products with this bacteria often go by the name "Bt."

Algae

Some products contain various kinds of soil-inhabiting algae. Most algae live in water, but a few can live in the upper part of moist soil. There they produce food by photosynthesis (adding to the nutrient supply of the soil ecosystem and stimulating other soil microbes) and release sticky substances (polysaccharides) that improve soil structure by gluing particles together. Some are classified as green algae and others are blue-green algae. The blue-green algae also can fix atmospheric nitrogen, as much as

25 to over 100 pounds per acre annually, according to company literature. Other benefits from algal products include improved root growth, increased yields and crop quality, increased water-use efficiency, reduced salt levels and less soil compaction (as deep as eight or 10 inches). The polysaccharides they release, plus increased root growth, can increase soil organic matter. A population of soil algae is something like interplanting a cover crop. They cover the surface, reduce crusting and protect from erosion. Some tests have found that inoculated algae do not colonize soil which quickly dries out, and that their polysaccharides may not reach much below two millimeters (unless tillage is done to incorporate them). In years with adequate rainfall, soil algae usually colonize fields naturally.

Kelp (Seaweed)

Kelp is a name for certain kinds of large "seaweeds," or marine algae in the brown algae group. Some species grow over 100 feet long in the ocean. The main species used for soil application is *Ascophyllum nodosum*, a medium-sized kelp that grows on rocks near the North Atlantic shores of Maine, Canada, Iceland and Norway.

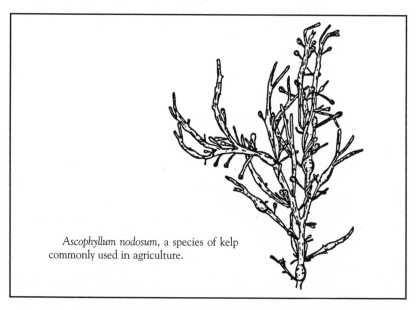

Ascophyllum nodosum, a species of kelp commonly used in agriculture.

This kelp is harvested and dried, then ground into a meal or powder, which has a long shelf life. It can either be used dry (applied to the soil) or it can be liquified as a solution that can be sprayed on the soil or on plants. Kelp can also be used as a seed treat or a root dip, and it is often mixed with other materials, such as fish extract (see below). As it grows in the ocean, kelps contain a wide variety of mineral elements (at least 60 elements have been found). Because they are plants, they also contain much organic matter as well as enzymes, hormones and vitamins. The benefits of kelp on soil and crops are many:

1. Improved soil structure. Kelp's carbohydrate content glues together soil particles, plus it also stimulates soil microorganisms, which in turn improve soil structure.
2. Increased nutrient availability and uptake. Certain organic constituents and trace elements of kelp provide plant nutrients in chelated (more available) form. Root cellular activity is also stimulated.
3. Enhanced germination. Soaking seeds in dilute kelp extract (about one part liquid kelp to 100 or 200 parts water) before planting can increase speed and percentage of germination. Kelp's hormone content is believed responsible.
4. Increased plant growth. Both roots and tops of plants have shown increased growth when proper amounts of kelp are used (as with any hormone material, too much can have a negative effect).
5. Increased yield or fruit set. Yield increases of 15 percent to well over 100 percent have been reported for many crops; however, other tests have found no significant effect. As with any biological product, results are often much better in poor soil or under stress conditions than in good soil.
6. Better crop quality. Increased nutrient content and longer storage life, along with increased frost and drought resistance, are often reported from using kelp.
7. Increased pest or disease resistance. Research on kelp has found that it can help control root-infesting nematodes. Crop resistance to aphids, spider mites and fungus diseases has also been reported.

Fish

Ever since the Indian Squanto taught the Pilgrims to put a fish in each hill of corn, the value of fish products for biological agriculture has been recognized. Fish wastes and unmarketable rough species are used, either from the ocean or fresh water. Fish products include fish meal (a solid material made of ground up fish) and "liquid fish," also called fish emulsion or fish solubles (a water-soluble extract from manufacturing fish meal). With a relatively high nitrogen content (about 10 percent for meal and 5 percent for liquid fish), plus 3 to 8 percent phosphate and up to 3 or 4 percent K_2O, along with secondary elements and trace elements, fish products can be a good biological fertilizer. Fish products do have a pronounced odor (unless a deodorizer is used), making you popular with neighborhood cats and unpopular with your family.

Fish materials are often combined with kelp to make a power-packed team, with the higher N-P-K content of fish complementing the trace elements, chelating and growth-promoting properties of kelp. From $\frac{1}{4}$ to $\frac{1}{3}$ of the mixture should be kelp and the rest fish. Either soil application or foliar sprays can be used.

Humic Acids, Humates

Humic acids are organic acids that occur in humus and other organically derived materials, such as peat and certain soft coal (or near-coal) deposits. There are different related varieties of humic acids: humic acid proper, fulvic acid and ulmic acid. These are a poorly understood group of organic substances, but they have many benefits for soil and growing plants. Good soil with adequate humus should have plenty of humic acids, and compost is an excellent way to add some. Poor soil can often benefit by adding a humic acid containing product.

Deposits of brown coal and near-coal (called leonardite or humates), plus rock layers containing high organic matter, are mined and ground up for agricultural use. They may also have benefits by supplying trace elements and by conditioning the soil (improving structure). These materials can be spread on the soil as is, or the humic acids from these deposits can be extracted with alkaline solutions to give a liquified humic acid product, for either soil or foliar spraying.

Because they came from plant matter, humic acid containing products have some of the same biological effects as kelp and plant hormones (covered earlier in this chapter). These include:

1. Improved soil structure. The humic and fulvic acid molecules bond together clay particles, giving rise to good, crumbly soil structure. Humic substances also stimulate soil microorganisms, which also improve soil structure.
2. Increased nutrient availability and uptake. Humic acids have chelating properties, thus increasing the plant availability of many nutrients. Humic substances also appear to stimulate root cellular activity.
3. Increased germination. Soaking seeds in a weak humic acid solution can speed up germination, possibly from increased water uptake and cell enzyme activity.
4. Increased root and top growth. Increased growth rates of from 25 to over 100 percent have been reported. Humic substances seem to have an effect similar to plant auxin hormones. As with similar growth-stimulants, too much can have no result or a negative effect.

5. Increased yield. Some reported yield increases run from 12 to over 100 percent. Again, results are usually better in poor soil or under stress conditions than in good soil.
6. Better crop quality. Increased nutrient content and longer storage life have been reported.
7. Increased pest or disease resistance. A few reports have been published of crop resistance to fungus diseases and insect pests.

As a conclusion, we can say that biological products *can* be useful tools in a crop production system. They can't substitute for a complete fertilizer program, nor can they substitute for good tillage and other management methods. They are not a magic wand. The place to begin is to correct problems with soil structure and the balance of major soil elements, but biologicals can help pull a crop through in a stressful year or give an extra boost to crop health and yield.

In my years of farming and research, I have used many "biological" products. Sometimes I think I see some response; then I try to repeat it and nothing — at least nothing we can measure or see. To use and spend money on these biological products *instead* of balancing fertility, proper tillage, green manure crops, etc., I think is a big mistake. The approach I presently use is to fit some of these products into my program as I would use a probiotic or kelp for livestock. For example, in my crop fertilizer blends that are row-applied, I always include a kelp-fish meal-humate blend for feeding the soil life, stimulating roots and providing some trace elements. The humate product works like a sponge to absorb some of the extra free minerals in the soil for later use. These materials have been fairly well researched, require no extra trips to apply and don't add a great deal of cost to the crop fertilizer program. I feel confident that most years they give me a good return.

Another product I like to use is liquid humic acid. I have been to enough farms, seen enough responses and read enough research reports to again feel confident on how to use it and believe that it is providing benefits. Any time farmers use a liquid plant food, put on liquid nitrogen and use herbicides, they should add humic acid. I like using the humic acid for its chelating or sponge-like properties, holding the extra free soluble fertilizer for future use.

As for herbicides, we are convinced you can reduce your rates of chemicals, provided the herbicide is applied at the proper time and is the right one for the job. As for amount of humic acid and cost, on a 12 percent solution, I recommend at least one quart each time I apply one of the above liquid mixes. If the quart costs $5.00, I would make up all or most of that expense by reducing the liquid fertilizer and nitrogen rate and herbicide. If you are presently getting poor weed control and your crops aren't doing what they should, I don't suggest you start by cutting rates. Evaluate the system and make the necessary changes and corrections.

I suggest the use of liquid fertilizers not as a fertility program, but as a pop-up, starter or foliar plant food. They are a part of a program, not a complete program. They would be too expensive for providing all the crop's needs and/or to correct soil fertility. The use of liquids is a convenience and a tool for emergency bail-out, as with foliars. Adding humic acid to these mixes improves their performance, plus gives added benefits.

If you have been or are using a biological product and really believe it is doing you some good, then use it. I have not heard of many reports of these products (when properly used) hurting the soil or crops. It is like taking my kelp pill every day. It may not be helping me, but I'm sure it's not hurting me, and it just feels right doing it.

Remember, these are add-ons, not complete programs. If things are not going well now and money is short, this is probably not a good place to spend or the practice to add. That is how these products get called "Foo-Foo Dust" or "silver bullets." Someone was hoping for a miracle or shortcut to doing the job right.

17

TILLAGE

Tillage refers to various operations in which the soil is worked or stirred. Tillage may be done for various reasons, and you should know what you are trying to accomplish *before* you start the tractor.

Why Till?

The basic reasons we till the soil are to manage:

1. Soil aeration — Without the presence of soil air (oxygen), crop failure is guaranteed.
2. Soil water — The soil needs to take it in when it rains, hold it without pushing out all the air, and bring it up to roots when the soil dries out.
3. Crop residues (or other organic matter)— In order to get them rotted down and the nutrients they contain released for the next crop — fast.
4. Soil fertility — Soils need a balance and mix of nutrients throughout the root zone. Water movement in soil also moves nutrients. Proper tillage mixes and distributes nutrients throughout the soil.

Types of Tillage

We can recognize four types of tillage:

1. Subsoiling — a deep tillage to fracture a compaction layer. Sometimes there is also a lifting action to loosen deep soil for better water management and deep root growth.

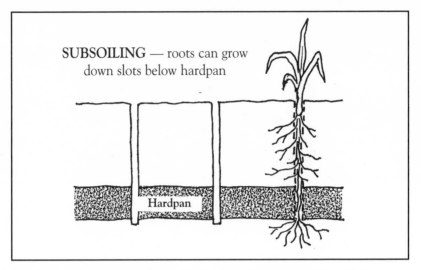

SUBSOILING — roots can grow down slots below hardpan

Hardpan

2. Primary tillage — a major mixing of the top six to seven inches of soil, to mix nutrients and organic matter. Each type of soil may need different kinds of primary tillage at different times.

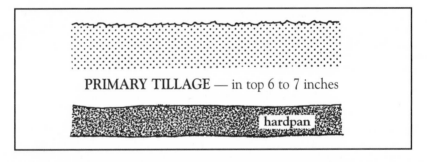

PRIMARY TILLAGE — in top 6 to 7 inches

hardpan

3. Seedbed preparation — this will be different for different crops. You need a good firm seedbed with ideal soil moisture and good seed contact.

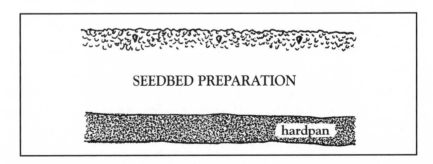

SEEDBED PREPARATION

hardpan

4. Cultivation — can be used for soil aeration, weed control, surface-mulching, hilling the crop or to side dress nitrogen.

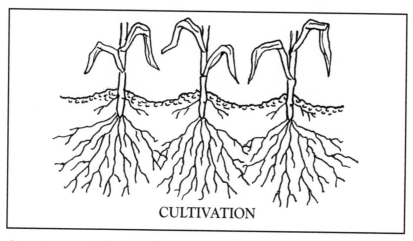

CULTIVATION

Cautions and Reminders

Soil compaction and *improper tillage* are two major limiting factors on many farms. You need to change your focus toward life in the soil:
• How do I get it?
• How do I keep it?
• How do I make the most of it?

To get maximum biological activity and a proper air-water-soil ratio, you need to remember these points:

1. In the beginning, mechanical aeration is essential. A fence post rots off at the surface, where oxygen is present, not a foot underground. A farm with high biological activity (at least 25 earthworms per cubic foot is a good measure) needs much less tillage because the soil life does it for you. But, you must prime the pump.
2. To prime the pump, you probably need to use all four types of tillage:
   a. Do subsoil in most cases. There are advantages.
   b. Don't allow tillage to deeply bury the food supply of soil life (the crop residues and manures).
   c. If a lot of fertilizers or manures have been used, a mixing is essential to avoid "nutrient stratification" (a strong concentration of salts in a narrow zone).
   d. Start at the top with proper nutrient use, and mix in lightly.
   e. Don't over-till. It can kill earthworms and deplete organic matter.

3. Don't lose your focus if your tillage method is not giving results. Every farm and soil is different. What Joe uses in central Illinois will probably not be right for Bob in Green Bay, Wisconsin.
4. Don't only evaluate a tillage tool by looking at short-term yields. What did it do for biological activity?
5. Tillage is only one part of success. You also need to follow the other rules of biological farming (see Chapter 2).

Many of the points in this chapter came from the book *Tillage in Transition*, by Don Schriefer. I strongly suggest you read it. Don works with a lot of corn and bean farmers and has studied soil physical conditions extensively. In his book he lists six "tillage commandments," which are worth noting:

• A tillage system must not be allowed to place limits on crop yields.
• A tillage system must address spring compaction and move toward elimination of all pre-plant spring tillage operations.
• A tillage system must guarantee a conditioned seedbed to provide a good start and uniform stands year after year.
• A tillage system must manage soil aeration, water and crop residue so as to nurture soil life and conserve and build the soil system.
• Every tillage operation must be done to remove one or more yield-limiting factors.
• A tillage system must address the potential problem of nutrient stratification and keep it from limiting yields.

To Till or Not to Till?

In the last two decades, reduced tillage systems have been much promoted, partly to reduce fuel and other inputs, and partly to reduce soil erosion. It might be called conservation tillage, minimum tillage or no-till. The idea is to have little or no disturbance of the soil, to leave crop residues on the surface as a mulch (to reduce erosion and conserve moisture), to plant through the surface mulch (requiring a new planter), to control weeds with herbicides rather than cultivation (higher rates are needed, however) and to apply lime and fertilizers on the surface.

Some of these practices do have benefits — but there are disadvantages also. Leaving residues on top does reduce erosion and does conserve moisture, but on the surface, much of the organic matter oxidizes into the air, wasting potential nutrients and humus. Sometimes too much moisture is conserved, and soil stays wet and cold, inviting fungus diseases and root rot. Lack of tillage and increased surface residues usually result in higher earthworm populations (as long as herbicides or pesticides don't harm them), but a lot of surface residue can also harbor over-wintering insect pests. Less field traffic and more earthworms decrease compaction, and less labor and fuel are used. Sometimes poor germination and low plant populations result from less than ideal seedbed preparation and possibly root diseases. In lighter, loamy soils no-till usually works pretty well, but in heavy soils in cooler climates, it can be a disaster. Soil temperatures can be as much as 10° F lower one inch below the surface. Surface-applied lime and fertilizer can be very slow to reach the root zone, so higher rates may be needed.

I like to see at least some tillage or soil mixing, because it lets air into the soil (giving faster spring warming and stimulating roots and soil life). It gets fertilizers and organic matter down where they can become available to roots. In no-till, root growth is usually shallow, sometimes resulting in lodging. Not all residues should be buried. Leave a little on the surface to reduce erosion. Mechanical weed control saves on expensive herbicides and aerates the soil at the same time. Moderate tillage done at the right time does not seriously harm earthworm populations. There are some reduced tillage systems that use occasional tillage, and with the right type of soil, they can be made to work. But, to be totally no-till without ever performing any tillage causes a lot of problems that can be easily prevented with wise, moderate tillage.

Choices

1. Growing a crop in a "chemical bath" of nutrients (mostly N-P-K) as in conventional agriculture . . . or

2. Growing crops in decaying organic matter in a soil with balanced nutrition and good biological activity.

Which do you think will do the best? Which will cause the least problems (weeds, insects, disease)? Which will perform under adverse conditions year in and year out? And which will be the most profitable?

When a crop is grown in decomposed green manure or livestock manure, or when a corn crop is planted in decaying soybean roots, it performs as if it were planted in freshly broken ground. I believe you can create this situation on your farm every year without trying to find fresh ground to break. Breaking fresh ground is and was the "old" way of farming. There is a better way.

Compaction

One of the worst problems involving tillage (as well as from other causes) is soil compaction. What is compaction? It involves an increase in soil density, along with a decrease in pore space. Tiny soil particles move closer together, and there is less room for air and water to move through the soil. Larger clumps of particles (soil aggregates, or crumbs) break down, allowing tiny particles to become cemented together or to be held tightly together by electrostatic forces. Compaction can occur as surface crusting, or as a hardpan in the upper layers or in the subsoil.

Among the problems caused by compaction are (1) low water infiltration and possible increased run-off and erosion; (2) difficult tillage (requiring more fuel); (3) poor root growth (from low oxygen levels and difficulty of root penetration) which can lead to poor crop growth in drought; and (4) low populations of beneficial soil organisms, most of which require oxygen.

The main cause of soil compaction are the mechanical forces of tillage and driving on the soil, *especially on wet soil.* Soil particles can move together much more readily when the soil is wet. Driving on wet soil causes compaction in both topsoil and deeper layers, while most tillage equipment tends to cause a hardpan to form at deeper levels.

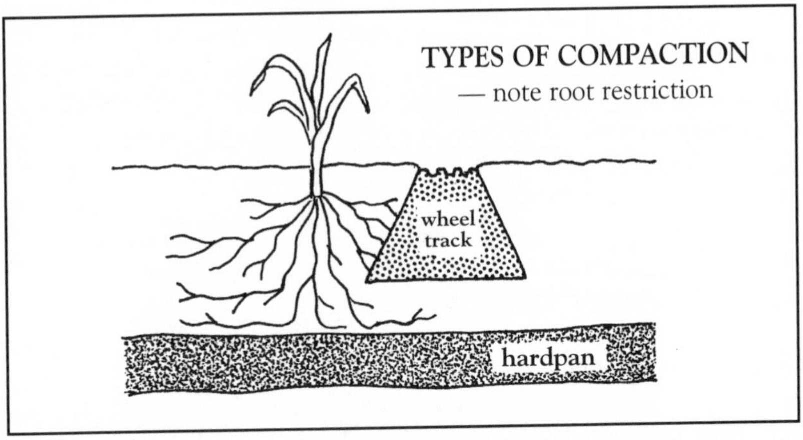
TYPES OF COMPACTION
— note root restriction
wheel
track
hardpan

Another major reason for compaction is low organic matter content of the soil, especially low humus. This usually is responsible for crusting as well as compaction in the upper several inches (the plow layer). Humus, plus sticky substances produced by roots and soil microbes, glues tiny soil particles together to form the larger aggregates which are necessary for good soil structure, aeration and root growth.

Other factors that contribute to soil compaction are uniform texture (all soil particles about the same size, rather than a mixture of sand, silt and clay particles), and the cementing or electrostatic attraction of particles, perhaps caused by excess salts or an imbalance in nutrient ions (too much magnesium or sodium and too little calcium).

Soil compaction can be reduced or overcome by the following methods:

1. Reduce the amount of driving (or only drive in the same tracks) and harmful tillage, especially on wet soil. Too early field work can really hurt crop growth and yields later on. Too much tillage reduces humus levels.
2. Increase soil organic matter (humus) by recycling crop residues, animal manures and green manures. Adding compost is excellent, too.
3. Plant cover crops or interseed grasses whose fine fibrous root systems greatly improve soil structure.
4. Encourage soil life, especially earthworms. Microbes and earthworms decompose raw organic matter to form humus. The burrowing of earthworms increases soil aeration and aggregation, and they can even break up a hardpan.
5. Avoid harmful fertilizers and toxic chemicals that destroy humus and soil organisms. Anhydrous ammonia, high-salt fertilizers and most insecticides and fungicides are the worst offenders. Balanced nutrients, with adequate levels of calcium, usually improve soil aggregation.
6. Address existing compaction problems now while waiting for the above methods to operate. Tillage such as rotary hoeing or cultivation for weeds will also break up a crust. Subsoiling or other deep tillage will break up a hardpan, allowing better water infiltration and aeration, and thus triggering some of the natural remedies.

To evaluate any tillage system, you need to dig. Get the shovel out, check root development, count earthworms, check the decay cycle and dig in the wheel tracks. Are you satisfied with what you see? Are there erosion problems? Is the soil loose and crumbly? These are all clues to evaluating tillage systems.

18

RECYCLING ORGANIC MATTER

Rule six in "How to Get Started in Biological Farming" (Chapter 2) is *feed soil life*. That does not mean putting on insecticides, herbicides, anhydrous ammonia, high-salt fertilizers or heavy coats of animal manure. To some degree, all of those kill beneficial soil organisms, or at least upset the delicate "soil ecosystem," the complex set of interrelationships among the many species of soil life discussed in Chapter 3. The organisms are in the soil to decompose raw organic matter and change it into *humus*, which has many benefits to the soil and to the crop (see Chapter 4). It makes no sense to purposely cripple this vital part of your soil.

Goals

To wisely manage organic matter recycling, you need to keep in mind what should happen:

- Raw organic matter (animal manures, green manures, crop residues or other materials) should decompose quickly so the nutrients they contain can become available to crops, and so humus can be formed.
- To decompose quickly, raw organic matter needs to be incorporated into the *upper* part of the soil, or the *aerobic zone* (which should be well aerated with a loose texture). You can find how deep your soil's aerobic zone is by seeing how deep the fine feeder roots grow. If you

have a good earthworm population, they will pull surface-applied organic matter into their burrows, but some kind of mechanical incorporation is more reliable.

- The soil should be moderately moist (not too wet and not dry) and warm (little decay happens over winter).
- Rates of application of raw organic matter should be moderate to low, especially for animal manures. Too much manure can ferment and release toxic by-products. Earthworms may be killed also.
- Remember that it takes some time for organic matter to decompose (approximately two weeks in the best of conditions). During decomposition, soil microbes can "compete" with growing crops for nutrients, and possibly some toxic by-products can be produced. Therefore, do not work in raw organic matter too close to the time for planting a crop.

Manure Management

I believe the best way to farm is to raise animals, so that manure is available for recycling. Without animal manures, it is more difficult and more expensive to provide the necessary nitrogen for crops and to build up humus levels. Growing legumes and incorporating green manures does help, but animal manures really can make biological farming work better — if they are managed properly.

Animal manures are potentially excellent sources of plant nutrients and soil organic matter, and they should not be thought of as a "waste problem." But improper manure management can result in wasted nutrients, damaged crops and polluted water.

Manure contains undigested nutrients from the animals' feed, plus excreted wastes from the kidneys and a considerable amount of bacteria (living and dead) from the intestine. In some systems, there may also be bedding mixed in. Animal manures have relatively more nitrogen than plant-derived organic matter. Here are some nutrient figures for *fresh* manure (from the University of Wisconsin Extension publication A1672, *Manage Manure For Its Value*, and other sources):

Source	% Moisture	Nitrogen pounds/ton	Phosphate (P_2O_5) pounds/ton	Potash (K_2O) pounds/ton
Dairy, solid	79	11	5	12
Dairy, liquid	92	5	2	5
Beef/Steer	74	14	11	14
Swine, solid	75	10	7	13

Source	% Moisture	Nitrogen pounds/ton	Phosphate (P$_2$O$_5$) pounds/ton	Potash (K$_2$O) pounds/ton
Swine, liquid	97	2	1	2
Horse, solid	65	14	5	14
Sheep, solid	65	21	7	19
Chicken, solid	75	25	25	12
Chicken, liquid	98	10	7	3

Source (solid)	Calcium	Magnesium	Sulfur	Iron	Copper	Zinc	Boron
			(pounds per ton)				
Dairy	5.0	2.0	1.5	0.1	0.01	0.04	0.01
Beef	2.4	2.0	1.7	0.1	0.03	0.03	0.03
Swine	11.4	1.6	2.7	0.6	0.04	0.12	0.09
Horse	15.7	2.8	1.4	0.3	0.01	0.03	0.03
Sheep	11.7	3.7	1.8	0.3	0.01	0.05	0.02
Chicken	36.0	6.0	3.2	2.3	0.01	0.01	0.01

These are average figures. You may want to have your own farm's manure tested to get more exact figures. The nutrient content of manure also varies at different times of year and with different kinds of feed.

The nutrients in manure are not all available to plants the first year. To calculate manure nutrient credits, you need to reduce the total nutrients in the above tables by a certain percent. Here are estimated first year nutrient availability figures (from the University of Wisconsin Extension publication A3411, *Manure Nutrient Credit Worksheet*; figures include nutrient losses, such as to air or in runoff).

NUTRIENT	DAIRY	BEEF	SWINE	CHICKEN
Nitrogen	35%	30%	40%	40%
Phosphate (P$_2$O$_5$)	55%	55%	55%	55%
Potash (K$_2$O)	75%	75%	75%	75%
Sulfur, Calcium, Magnesium	55%	55%	55%	55%
Trace elements	65%	65%	65%	65%

Much less of the nutrients becomes available during the second and third years the manure is in the soil; roughly 5 to 13 percent of the nitrogen in the second year and 2 to 8 percent in the third year. Of course, nutrient availability also would be higher in aerated biologically active soil than in compact, "dead" soil.

Manure nutrients are sometimes calculated on the basis of 1,000 pound animal units. For example, a 1,200 pound cow would equal 1.2 animal units; or 20 hogs weighing 275 pounds would be 20 X 275 divided by 1,000 = 5.5 animal units. For young growing stock, an average weight for their growth period can be used. The following table gives estimated annual nutrients per 1,000 pounds animal unit (from *Wisconsin Agriculturist*, Jan. 12, 1980, p. 34) (for less than one year, use a daily rate).

SOURCE	NITROGEN, lbs.	PHOSPHATE, lbs.	POTASH, lbs.
Dairy, solid	91	50	112
Beef, solid	77	77	99
Swine, solid	102	107	124
Swine, liquid	95	111	119
Chicken, solid	306	332	136

But not all the nutrients in manure may reach your crops. There are several ways some or even most of them can be lost — wasted — and they can pollute the environment as well. Some losses occur almost immediately after the manure is produced, while other losses may occur months later in the soil. Nutrient losses can occur from:

• Loss of nitrogen in the form of ammonia gas from bacterial action, drying, freezing, high pH (alkaline) and high temperatures. About 20 percent of manure nitrogen can be lost in only six hours, 25 percent in 24 hours and 45 percent in four days. Adding a nitrogen-conserving material to fresh manure as soon as practical will greatly improve the final nutrient value in the field, as well as cut down on odors. Ground rock phosphate is an excellent manure additive, as is ordinary superphosphate (0-20-0) and gypsum (a lot of lime is not good because it raises the pH). Another approach is to use a bacterial manure inoculant, which will speed up decomposition and tie up nitrogen.

• Loss of all nutrients by run-off from stacked manure or manure spread on the surface of the ground or on snow or ice. Manure should be worked into the soil as soon as possible.

• Loss of nitrogen in the form of nitrate by leaching from the soil. Either too much manure was applied for crop needs, or crop use was too slow. For fall application, planting a grass cover crop, or catch crop (such as rye), that will tie up nitrogen, is a good practice.

Here are some other manure management tips:

- A light coat on many acres is better than too much on a little land. This is the most important thing to remember.
- Spread evenly and incorporate into the top several inches, as soon as possible. Knifing in liquid manure deeply is a bad practice because it cannot decompose properly, and much nitrate can leach into the groundwater.
- Subsoiling and/or soil aeration before applying manure reduces losses.
- Fall application seems ideal in most cases; that way it has time to decompose before planting.
- On low-testing (phosphorus and potassium) fields, use up to 12,000 gallons/acre of liquid manure; no more than 6,000 gallons/acre on medium-testing fields and no more than 3,000 gallons/acre on high-testing fields (1,000 gallons of liquid manure = four tons of solid manure).
- Never apply manure to hay fields during the growing season. Only apply it to low-testing fields in the fall, and never over 3,000 gallons/acre of liquid manure.
- Spreading in cool, rainy weather will reduce nitrogen loss (but not in heavy rain or there will be run-off).
- One main reason to have soil tested from all over the farm is to evaluate where to spread the manure and how much to apply. Distribute manure to bring all fields to the same level of fertility. This does wonders for a balanced, uniform feed supply.

Green Manures

Green manuring means adding fresh, green plants to the soil. It is thus different from adding dried crop residues, animal manures, compost or any other organic matter. Usually the fresh plant matter is incorporated into the upper layers of the soil, but in some cases it may be left on the surface.

Green manuring is an excellent practice and can greatly improve the soil and stimulate crop growth. It increases the population of beneficial soil organisms by giving them a food supply. Any kind of organic matter will increase soil organisms, but the fresh plant tissues of green manures contain more available sugars, amino acids, enzymes, vitamins and other cell materials than dried residues, so there is an almost instant "population explosion" of soil microbes.

Adding organic matter to the soil can (1) increase the humus level, and (2) increase the nitrogen level. Most plant matter doesn't do both at once very well. Young succulent plants tend to raise the nitrogen level, but because they have little fiber, do not add much humus. Mature, coarse plants will increase humus, but they may even deplete nitrogen as they decompose. Succulent plants decompose rapidly and cause a quick burst of microbial activity, while coarser plants decompose more slowly. Legumes generally add more nitrogen than non-legume species because they are high in protein and usually have nitrogen-fixing root nodules. Deep-rooted plants are valuable since their roots can bring up minerals from the subsoil.

Green manure crops, while they are growing, can also serve as catch crops, absorbing and holding nutrients that might otherwise leach away, such as nitrogen. The nutrients will later be released when the plant decomposes. A growing green manure crop can also act as a cover crop, protecting the soil from erosion as well as smothering weeds.

If your goal is to increase humus, don't forget that you will be adding both the tops and roots. A plant's root system can make up $1/3$ to $1/2$ of its volume. As a plant grows, new roots are constantly growing and old ones are dying and adding humus to the soil. Grasses are especially valuable because of their extensive fibrous root system, which can literally have hundreds of miles of roots on a single plant.

There are different ways to use green manure crops:

- If you have loamy soil with good structure, you will get the most benefit from growing a legume and incorporating it when it is young. It will add nitrogen and stimulate soil life.
- If you have dense clay-like soil, perhaps with a hardpan, or light sandy soil, you would benefit most from added humus. Humus will loosen dense soils and hardpans, and improve the structure and fertility of sandy soils. Green manures are sometimes left on the surface of sandy soils, to keep nutrients near the top (because sandy soils leach readily). However, green manures seldom work well in dry climates because there is not enough soil moisture for them to decompose properly.
- To add humus, a grass might do well, or else a mixture of grass and legumes. With a grass-legume mixture, each plant contributes its specialty: the legume adds nitrogen and grass roots improve soil structure. Choose species adapted to your climate.

Unless you are starting with poor soil or have a set-aside field, it is not usually economical to take cropland out of production just to grow a green manure crop. The most common practice is to plant a green manure crop in the late summer or fall, use it as a ground cover over winter (to reduce erosion and to shelter earthworm populations from sudden frosts), and then incorporate it in the spring. Some species or varieties may winter-kill in cold climates. In warm climates, a fast-growing green manure crop can be grown between two successive high-value crops, such as vegetables. A green manure can also be interseeded into a row crop and later be incorporated along with the residues.

The crop following the green manure crop should be planted soon enough after incorporating the green manure to use the released nutrients (before they leach away), but not too soon. The decomposing plant matter will temporarily tie up some nutrients (especially nitrogen) and thus compete with (and possibly stunt) an emerging crop. In the spring, wait at least two weeks after incorporating a succulent green manure crop before planting another crop. For a coarse green manure crop or in cool weather, it may take a month. If leaving the soil alone for this time period is not possible, be sure to use a starter row fertilizer. I use row cleaners on my planter. They remove the decaying residues and allow me to plant in a fresh, clean soil, which gives good soil-seed contact.

Some crops that do especially well after a green manure crop include corn, potatoes, tobacco, rye and oats. Wheat and barley may or may not be helped. In general, crops that need a shot of nitrogen do well.

There are many species of legumes and non-legumes that make good green manures. Here are some that I have seen, and typical seeding rates:
• Oats — two to three bushels/acre.
• Ryegrass (fall seeded) — 1.5 to two bushels/acre.
• Annual ryegrass (mixed with clovers, interseeded in corn) — 10 pounds/acre each.
• Clovers (interseeded in corn) — 10 pounds/acre.
• Vetch — 15 to 30 pounds/acre.
• Buckwheat — 30 to 75 pounds/acre.
• Brassicas (rape, turnip, oilseed radish; good for late fall growth) — two to four pounds/acre.

Composting
Composting is the process of allowing fresh organic matter to decompose into humus. When manure is spread and incorporated into the soil, it decomposes. This is sometimes called *sheet composting*. However, the usual concept of composting is to let organic matter decompose before it is added to the soil.

Composting takes extra time and trouble, but it can have several advantages:

Manure and other organic matter is turned into humus before going into the soil, eliminating possible toxicity from too much manure or anaerobic decomposition.

- Compost is usually a balanced, natural fertilizer, with plant growth-stimulating substances.
- Compost inoculates the soil with beneficial microorganisms.
- Compost improves soil structure, aeration and water-holding properties.
- Composting destroys weed seeds and pathogenic microorganisms.
- Composting reduces the volume of the material to 20 to 75 percent of original; thus there is less material to haul to the field.
- Composting reduces odors when spreading.
- Compost is in demand by gardeners and organic growers; you may be able to sell what you don't use.
  Here are some composting tips:
- You need to provide a good home for the "volunteer army" of microorganisms that decompose the raw organic matter. They need (a) food (organic matter), (b) moisture, (c) air, (d) proper temperature and, (e) absence of toxic substances.
- Almost any fresh organic matter can be composted, including animal manure, green manure, crop residues, straw, old hay, weeds, city leaves and grass clippings, tree bark, sawdust and wood chips, brewery and cannery wastes, soybean and cottonseed meals, hulls and shells, paper mill and sewage sludges, shredded newspapers, garbage, slaughterhouse wastes (bonemeal, bloodmeal, feathers, hoof and horn meal, tankage) and fish meal.
- A "balanced diet" for microbes will have some high-nitrogen material (animal manures or other animal by-products, or legumes) as well as high-cellulose material (most plant matter except legumes). Best composting occurs when there is a mixture of about $\frac{1}{3}$ high-nitrogen and $\frac{2}{3}$ high-cellulose materials. Therefore, straight manure is not as good as manure with bedding or other plant matter mixed in.
- The raw organic matter should be piled in such a way that a fairly large surface area is exposed. For agricultural composting, long windrows about four to five feet high and eight to 12 feet wide at the bottom work well.
- The material should be kept moist inside the pile, but not waterlogged. Add water if it begins to dry out. A moisture content of 40 to 60 percent is ideal, although composting will occur from 25 to 75 percent moisture. If you squeeze a handful of the material with your fingers, it should hold together but no water should squeeze out.
- The pile should be turned (mixed) one to several times during composting to help aerate it, cool it, dry it out (if too wet), or to better mix the different materials. Turning the pile speeds the rate of decom-

position (unless the temperature is too low, below 85° F). A front-end loader can be used to turn a pile, or it can be loaded on a manure spreader and re-piled. Special compost-turning machines can be purchased or built.

- When other conditions are favorable, the temperature inside the pile will rise. Temperatures from about 100° to 150° F are ideal. If the temperature inside the pile gets above 160° F, the pile should be turned to cool it off.

- Adding about one-forth soil to the pile will help supply microorganisms, absorb ammonia and speed composting. Adding a little lime or rock phosphate as a calcium source will stimulate microbial activity. The pH inside the pile should not become too acid or too alkaline (don't add a lot of lime), or proper decomposition will not occur.

- If composting does not occur rapidly, adding a special microbial "starter" product may help. Some previously made compost makes a good starter, as does "compost tea" made by soaking some compost or rich soil in water for several days. If composting still does not occur, check the other factors (moisture, temperature, aeration, balance of raw materials).

- Depending on weather (mostly temperature) and the other factors, it will take about three to 12 months for finished compost to be formed. It takes longer in cold weather than warm. There are special techniques that can produce finished compost more rapidly, in weeks or even days. Compost is "finished" when the inside temperature falls to near the outside temperature and there is no ammonia odor. Finished compost should be a dark, homogeneous material (no stems or leaves remaining) with a rich, "earthy" smell (no unpleasant odors). Finished compost should be kept moist and covered to protect it from rain which would leach its nutrients.

- Compost is essentially humus and can be put on soil at any time without harming crops (unless the crop doesn't need a lot of nitrogen, like small grains). It is like instant plant food. It is best to till it into the soil to keep its microbes active. Good results have been obtained at rates from $\frac{1}{2}$ to 10 tons per acre.

19

WEED CONTROL

A weed is often defined as a plant growing where we do not want it — usually where we are trying to grow crops or a lawn. Weeds generally cause some kind of harm or economic loss:

- Compete with crops for water.
- Compete with crops for soil nutrients.
- Compete with crops for light.
- Lower crop yields and quality.
- Contaminate grain with seeds.
- Harbor crop diseases or pests.
- Taint milk with unwanted tastes or odors.
- Harm livestock by poisoning or from thorns and spines.

In spite of all these bad things, weeds are not totally worthless. They have their place in nature. They quickly colonize bare soil and prevent erosion. Their roots help loosen hard soil and may bring up nutrients from the subsoil. They can scavenge and conserve excess nitrogen that might otherwise leach away. When they die, they add organic matter to the soil.

Some weeds even make nutritious animal forage (lambsquarters and pigweed are sometimes grown for forage in Europe). Weeds also provide food and shelter for wildlife. Some are useful as herbs or sources of medicine, as natural dyes and even as tasty "wild foods" for people. Finally, some weeds can teach us something about our soil and agricultural practices. Those weeds that prefer certain types of soil are called indicator species. For example, foxtail, velvetleaf, mustards (yellow rocket, etc.), fall panicum and crabgrass like hard or crusted soil with little air. On the other hand, lambs quarters and redroot pigweed generally indicate fertile soil.

Weed Facts

- There are about 1,775 species of weeds in the United States, including aquatic weeds.
- Weeds can serve as companion plants for garden crops:
  - corn with cocklebur
  - potato, carrot, tomato, radish, beet, pepper and eggplant with redroot pigweed, lambsquarters, sow thistle, bull thistle and goldenrod
- Weeds can produce vast amounts of seed; for example (per plant):
  - black mustard, green foxtail — 140,000
  - crabgrass — 200,000
  - lambs quarters — 600,000
  - tumbleweed — 6,000,000

- Weed seeds can germinate after amazing lengths of time:
  - St. Johnswort — 10 years
  - Canada thistle, yellow nutsedge — 20 years
  - common chickweed — 30 years
  - shepherd's purse — 35 years
  - redroot pigweed, Virginia pepperweed — 40 years
  - curled dock, black mustard, field bindweed — 50 years
- Weeds absorb a lot of nutrients. Compared to corn:
  - common ragweed uses 50 percent more nitrogen, 34 percent more phosphorus, 61 percent more potassium, 87 percent more calcium and 48 percent more magnesium
  - lambsquarters uses 53 percent more nitrogen, 76 percent more phosphorus, 73 percent more potassium, 88 percent more calcium and 72 percent more magnesium
- Weeds absorb a lot of water. Following are crop and weed figures for pounds of water needed to produce one pound of dry matter:
  - corn — 368
  - redroot pigweed — 287
  - wheat — 513
  - lambs quarters — 801
  - alfalfa — 831
  - ragweed — 948

The most troublesome weeds are foreigners — they originally were brought to this country from another part of the world, usually accidentally. Infestations of noxious weeds can be brought on by poor weather, wrong soil conditions (too wet, poor structure, etc.) or out-of-balance fertility (higher calcium usually reduces bad weeds). Rather than considering weeds the problem, look at them as symptoms of a problem. Conventional agriculture tends to ignore the message that weeds give us about soil conditions and just tries to wipe them out with herbicides — without fixing the problem that caused them to grow in the first place. Sure enough, the weeds are back again next year. Recent studies have found that constant use of herbicides can even *increase* weeds, and some weeds have become resistant to herbicides. If what you're doing isn't working, it's time to try something else.

Weed Control Principles

When we say "weed control," we do not mean eradication. Since the soil contains abundant weed seeds, some of which can sprout after 40 or 50 years, we will always have weeds to contend with. Most weed seeds sprout only when conditions are right for them, such as the low oxygen/high carbon dioxide conditions of compact or wet soils, or when they are stirred up to the surface (light can trigger them). Many weeds germinate at cooler temperatures than crops such as corn or soybeans, so they get a head start on the crop.

When you are deciding what to do to control weeds, you have to consider whether the weeds are going to cause a significant loss in crop yield or quality. Wiping out all weeds just so you can have a totally clean field is not an effective use of resources. Weed experts speak of an "economic threshold," or a certain level of weed infestation that will seriously harm crops. It varies with the weed and the crop. Usually you have to do weed counts several places in a field and calculate from a formula to see if control is worthwhile.

The most critical time to reduce weed pressure is early in the growing season, when the crop plants are trying to establish their root systems. A large root system is critical for drought resistance, better yields and crop quality. Crop seedlings are especially vulnerable to competition, and some weeds harm crops at this time by toxic substances they release. This "chemical warfare" among plants is called allelopathy.

One of the keys to effective weed control is timing. Eliminate the weeds when they are the most vulnerable, whether you are spraying herbicides or cultivating or anything else. The best time for weed control is as early as possible, when the weeds are just sprouting (in the "white-root stage"). With shallow roots, weed seedlings are easily killed, and they are eliminated before they can harm the crop.

For weeds that sprout in cool weather, sometimes it works well to let the early weeds germinate and then kill them while you are preparing the seedbed for corn or other row crops. Being the first in your neighborhood to get your crops planted may backfire if there is a cool spell and the crop stands still while weeds go wild.

If you are trying to control weeds that have already grown, be sure to kill them before they go to seed. Some weeds have low root reserves at flowering and seed production stages, so they are more easily killed by mowing or herbicides at that time.

For successful long-term warfare against weeds, it is wise to "scout the enemy." Survey your fields, both early and later in the season to see where the weed problems are. Try to correlate weedy spots with soil conditions (soil type or structure, drainage or organic matter) and with past practices (crops, tillage or fertilizers used). If a soil problem seems to be the cause, plan to do whatever is necessary to correct it, such as deep tillage to improve drainage, or reducing the amount of manure.

Herbicides

Man-made herbicides are chemicals designed to kill plants. Modern herbicides kill by different mechanisms, usually by interfering with the plant's cellular functions, such as by crippling photosynthesis, amino acid production, cell division, energy transfer, cell membrane functions or plant hormones. Some herbicides are more specific than others in exactly which plants they will kill. Some work only on grasses, while others kill broadleafs. Many herbicides are systemic, meaning that they are absorbed inside the plant's tissues. Most modern herbicides are supposed to degrade or break down fairly quickly after they are applied. Herbicides certainly *can* give effective weed control.

But herbicides have their down-side, too. Sometimes they don't work at all, such as in a too-wet or too-dry year. They may kill plants you don't want killed, such as instances of crop damage from herbicide carry-over, or your herbicides visiting the neighbor's crops on a windy day. We see an increasing amount of herbicide damage on the growing crop. Maybe it's weather related, maybe it is complex soil chemistry interaction, and maybe it is soil biological changes causing the damage. A few herbicides can kill beneficial soil organisms (earthworms, bacteria or fungi. Atrazine has a reported history of suppressing soil bacteria and Roundup seems to be interfering with soil fungi particularly soil mycorrhizae. The real truth is that not much is known. The soil life is a complex, poorly understood world. In most cases, we can honestly say we don't know what effect herbicides have on soil life. And as we have noted, some weeds have become resistant to herbicides.

Then there are the horror stories of possibly cancer-causing herbicides turning up in well water and groundwater across the nation. They didn't break down as fast as they were supposed to, or else their breakdown products are also toxic. And consumers worry about herbicide and pesticide residues in their food.

Then there is the extra expense of using herbicides, easily $50.00 or $60.00 an acre for some crops. Perhaps the most insidious negative effect of relying on herbicides for weed control is that it blinds you to finding out what the real soil problems are and fixing them. Herbicides are just a crutch, a band-aid. They can give some temporary relief, but they don't cure the sick patient.

A Better Approach

With all the economic and environmental problems created by the high-chemical methods of conventional agriculture, would it make more sense (and cents) to reduce unnecessary expenses, to *prevent* problems that rob your profits, and to grow high-yielding and high quality crops in a way that minimizes environmental damage?

That is exactly what a biological farming system does. By working with natural systems in the soil, weed pressure can be greatly reduced, and crop growth and quality increase also. By using better fertilizers, proper tillage and recycling of organic matter, the poor soil conditions that are responsible for most weeds can be eliminated. And by relying more on non-toxic weed control methods for most weed problems (and maybe using herbicides at reduced rates only as a back-up), chemical expenses and environmental pollution can be greatly reduced.

Non-Toxic Weed Control

There are many ways to control weeds other than using toxic herbicides. We can group them into four main categories.

Crop rotation

It is well known that monoculture, growing the same crop year after year, brings with it certain weed species. It can be related to tillage or fertilization, but another factor is degraded, low-humus soil from continuous row crops like corn, cotton or soybeans. Usually, as the years go by, higher rates of fertilizer are needed, along with pesticides and more herbicides.

Many scientific studies have proved that when crops are rotated, not only do yields improve (with less outside fertilizer being required), but soil structure improves and weed populations (species) change. Each type of weed grows best under certain soil conditions and nutrient balance. One study found 25 weeds per square yard in continuous corn, but only four in a corn-soybean rotation. Some research studies have found that although weed species change in a rotation, weed pressure may not. However, many biological farmers *have* found a definite decline in weed pressure after a few years, and many only use a little herbicide for spot spraying, or use none at all.

The likely reason the research scientists did not always get weed reduction from rotations is that they didn't pay attention to soil structure, nutrient balance, humus and soil life. They didn't fix the real problem.

Obviously, you have to plan a sensible crop rotation that works for your farm (for your soil, climate and pocketbook) and one that eventually builds up and improves the soil so the wrong soil conditions can be eliminated. You have to alternate soil-building crops (legumes, fine-rooted grasses) with soil-depleting crops (corn). Growing nitrogen-fixing legumes just before high-nitrogen-consuming crops like corn means less purchased nitrogen is needed. Hay crops are excellent soil builders. Even if you have no livestock, hay can be grown as a cash crop, or grass/legume cover crops can be planted in the fall between grain crops.

A short or "tight" rotation works better than leaving a field in the same crop for more than a year or two (or three for hay). The idea is to keep the weeds "off balance" and to have a different crop just about every year (or two). Also, some crops may suppress weed growth by shading or by allelopathic effects. Common allelopathic crops include rye, barley, oats, wheat, corn, tall fescue, sorghum, Sudan grass, soybean, alfalfa, red clover, pea, field bean, sunflower and buckwheat.

Improve Soil Conditions

Since the basic cause of most weed problems is wrong soil conditions, obviously correcting what is wrong with the soil will greatly reduce weed pressure (there could still be some occasional problems caused by bad weather, and there are a few weeds that grow in good soil). What many farmers have been doing unknowingly is fertilizing and tilling their fields to encourage weeds and discourage crops. They need to reverse that.

There are three main aspects of the soil that affect weed versus crop growth: soil structure, fertility level and nutrient balance, and soil organisms. The farmer needs to do whatever he can to make his crop plants grow better and weeds grow less well. A vigorously growing crop can shade out the weeds, and below ground a vigorous root system can out-compete the weeds for water and nutrients.

Soil structure. The ideal soil structure, which promotes crop growth and discourages many weeds from germinating or growing well, is a loose, granulated, well-drained soil. When the tiny soil particles are clumped together into larger "crumbs" (or aggregates), water will drain downward after a rain (eliminating waterlogging), air will enter (roots and beneficial soil organisms need oxygen), and roots will grow deeper and faster through the soil. Also, a high organic matter (humus) content (3 to 5 percent) provides good water-holding capacity and crop nutrients.

On the other hand, a surface crust cuts off the movement of air, and sub-surface compaction such as a hardpan prevents water drainage and deep root growth. Without granulation (good aggregate structure), the entire soil can be dense and compact. In dense soil without adequate aeration, recycled organic matter (crop residues, manures) do not decompose properly, but instead release toxic substances.

The seeds of many weeds, such as velvetleaf, foxtail and mustards, are "programmed" to mainly germinate when the soil is poorly aerated (low oxygen or high carbon dioxide), or is too wet or laced with the toxic by-products of wrong organic matter decomposition. If a field with low humus is over-tilled (which destroys soil structure) and planted with a crop, then when a rainstorm causes a surface crust . . . presto . . . up pop the foxtail and velvetleaf.

To discourage weeds that flourish in hard, tight, crusted or wet soil, you must do several things that will improve soil structure. First, break up any existing compaction, whether at the surface (by shallow tillage or cultivation) or deeper (by subsoiling or deep tillage). This will allow air and water to enter more freely and "prime the pump" for more improvement. Avoid causing additional compaction by not tilling or driving on wet soil.

The common practice of reduced tillage or no-till is sometimes helpful to weeds and harmful to crops. In coarser, lighter soil types, no-till can work well, but in fine-textured, heavier soils in cool climates, the upper several inches tend to become denser and poorly aerated — an ideal environment for many weeds to grow. Don't use no-till just because it is fashionable; check to see what it does to your soil.

Second, increase the soil's humus level (if it is below 5 percent organic matter on a standard soil test) by adding organic matter in moderate amounts into the upper few inches (not too much or too deep). In aerated soil, such materials as animal manure, crop residues, green manures (turned-under green plants), sewage sludge, slaughterhouse or cannery wastes, or compost will decay into humus (see Chapter 18). Humus will increase soil granulation and water-holding capacity, and also help nourish crop plants.

Third, grow crops that improve soil structure, either as a main crop (hay or forage) or as a cover crop or green manure. These plants improve soil structure both by adding organic matter when roots die and from sticky secretions of living roots. The best soil-improving crops include fine-rooted grasses (such as rye, ryegrasses, brome, fescue, etc.) and legumes (such as clovers, vetch, birdsfoot trefoil, alfalfa, etc.). A grass-legume mixture works very well because the nitrogen from the legume's root nodules helps the grass grow. For significant soil structure improvement, such crops should be grown for at least a couple years, but even a fall-to-spring cover crop helps a little.

Fourth, be sure there are no serious imbalances of elements that affect soil structure (see Chapter 6). For typical soils, the most abundant cation (positively charged) element should be calcium (from about 70 to 75 percent or more of exchangeable cations), followed by magnesium (12 to 15 percent) and potassium (3 to 5 percent). Hydrogen and sodium should be less than 5 percent. Soils with low calcium and high sodium or magnesium usually have poor structure. Using the proper kind of liming material or calcium source is very important in weed control (see Chapter 10). In areas with already high magnesium, using dolomitic lime (with 15 or 20 percent magnesium) does not help.

Using commercial fertilizers that are toxic or that cause nutrient imbalance can lead to poor soil structure and weed problems. In addition, highly soluble fertilizers leave excess soluble nutrients to fertilize the weeds. Types of fertilizers used and careful placement can greatly reduce the fertilization of weeds. Anhydrous ammonia and fertilizers that release high amounts of ammonia (solid urea and diammonium phosphate, or DAP) can kill crop seedlings and cause worsening soil conditions (acidity, loss of humus and denser soil). High-salt fertilizers such as muriate of potash (or Kalium potash) can also injure seedlings and cause salt build-up which may favor weeds.

Fifth, encourage a high earthworm population (see Chapter 5). With their tunneling and production of casts, earthworms can eliminate compaction and increase granulation of soil. They need a supply of fresh organic matter (manure, crop residues, etc.) for food and as little disturbance as possible from tillage (at least until their population builds up). Other beneficial soil life also helps improve soil structure — the microscopic bacteria and fungi that decompose organic matter. They need aerated soil and occasional additions of raw organic matter for food. Usually if earthworms are common, the microscopic organisms will be too. Earthworms often eat weed seeds and either destroy them or lower their germination capability. Some weed seeds may be destroyed by microorganisms. Composting manure or other materials is a good way to kill weed seeds.

Soil life. We have just mentioned the value of beneficial soil organisms in improving soil structure. Their production of humus along with sticky materials that they release, glues soil particles together into larger aggregates, which increases soil aeration and drainage.

Many soil organisms also indirectly help fight weeds by helping crop plants grow better. That way the crop can overtake and shade out the weeds. Microscopic bacteria and fungi that live on the roots' surface aid roots in absorbing water and nutrients. They also release growth-promoting substances such as vitamins and hormones. And they protect roots from invading diseases and pests. All of these benefits add up to a healthier, more vigorous crop plant.

Soil nutrients. Many weeds grow best in poor, infertile soil, perhaps with nutrient deficiencies or excesses, high salts, or too low or too high pH. High nitrogen favors some weeds, as does the high potassium that often results from too much manure. Crop plants grow best when all essential nutrients are available in amounts sufficient for that species — and in a proper balance (see Chapter 6). Many experiments have shown that crops grow and yield better and produce better quality food when all nutrients are available (not just N-P-K) and when they are in balance. Crops need adequate calcium, magnesium, sulfur, iron, manganese, copper, zinc and boron, besides nitrogen, phosphorus and potassium.

An excess of one element often causes a deficiency of another. Calcium, magnesium and potassium work that way. Soils naturally high in magnesium produce crops low in calcium (and such soils may have poor structure also). Using too much potassium fertilizer will lower the crop's intake of calcium and magnesium. But many weeds love high-potassium, low-calcium soils, such as burdock, dandelion, crabgrass, quackgrass and red sorrel.

Maintaining a high fertility, balanced soil with high calcium will eliminate a lot of troublesome weeds — they just don't want to grow in that kind of soil. Crops will grow better in balanced soil and will out-compete the weeds that are there (a few weeds do grow well in good soil, like lambsquarters, but they are relatively easy to control by other methods).

Keeping the soil pH near the "ideal" range of 6.2 to 6.8 is also helpful in fighting weeds. A few weeds grow best in very acid or very alkaline soil. But the main reason for keeping the right pH is that soil nutrients are more available to crops, and they are in a good balance when the pH is from 6.2 to 6.8. Beneficial soil organisms also do best in those pHs.

A good strategy to encourage crop plants and discourage weeds is to keep ideal soil conditions and fertility in the row but not between rows. Fertilizer can be banded in the row (that saves money, too — why fertilize weeds?), and an ideal seedbed can be prepared in the row while leaving the ground loose and cloddy between rows. Those conditions will slow down weed seed germination and hasten crop growth.

Grow Smother Crops and Cover Crops

Almost any plant will languish or die when it is shaded. It just can't get enough sunlight to make sufficient food to sustain itself. Gardeners and vegetable growers use this principle when they control weeds with a mulch of straw or other shade-producing material. Most weeds germinate and get a foothold on bare soil, the same situation that farmers create when they till and plant a crop.

An excellent approach to agriculture which helps control weeds *and* improves soil is to keep the soil covered year-round. A plant cover shelters the soil from erosion and from temperature extremes (it also shelters earthworms). Plants improve the soil (both structure and fertility) by the organic matter they add and the organisms they encourage. And densely growing plants can effectively shade out or crowd out weeds. They can also be turned under as green manures. Certain species can kill weeds by allelopathic effects (see the previous section on crop rotation). Crops grown to provide ground cover are called *cover* crops, and those planted specifically to control weeds are often called *smother* crops. The nurse crops for new seedings of legumes, such as oats or barley, also make good cover crops.

A cover or smother crop can be grown as a main crop in a rotation, such as a hay or forage crop. Perennial (several-year) species can effectively eliminate annual (one-year) weed species. Or, if you have no need for such crops, a temporary cover crop can be grown between the main crops in your rotation. Typically, the cover crop is planted in the late summer or fall, stays on over winter, and is turned under the next spring.

Another method that provides ground cover and weed control in row crops is interseeding the cover crop between the rows (overseeding by broadcasting also works well). If the main crop's canopy provides a lot of shade, the cover crop may sprout, but then "stand still" until the row crop is harvested. Then the cover crop will grow more rapidly and be there for the winter.

Some good cover and smother crops include (1) legumes such as clovers, vetch, birdsfoot trefoil, alfalfa, peas, soybeans, annual medics and lespedeza; (2) grasses such as rye, ryegrass, barley, oats, winter wheat, sorghum or sudan, millet, reed canary grass, crested wheatgrass and fescue; (3) corn for silage; (4) buckwheat; (5) forage brassicas such as turnip, rape and kale); and (6) sunflower. Also good are mixtures, including rye-vetch, oats-rye, alfalfa-brome, alfalfa-red clover, alfalfa-sweet clover-red clover, red clover-annual ryegrass, and yellow sweet clover-rye.

COMMON COVER, SMOTHER & GREEN MANURE CROPS

Crop	Planting	Seed lbs/acre*	Comments
Legumes (for best growth, inoculate seed with proper nitrogen-fixing bacteria)			
alfalfa	spring	15-20	Adds much nitrogen. Also valuable as hay, silage or green chop.
alsike clover	spring/late summer	3-10	Tolerates wet or acid soil. Good with a grass, such as redtop.
berseem clover	spring-fall	10-20	Annual, does not overwinter. Tolerates wet soil. Good forage or pasture.
birdsfoot trefoil	spring	15	Tolerates poor soil. Good as a forage.

COMMON COVER, SMOTHER & GREEN MANURE CROPS
(continued)

Crop	Planting	Seed lbs/acre*	Comments
cowpea and field pea	spring	20-100	Does not overwinter. Tolerates poor soil Good with oats. Good forage.
crimson clover	spring (N) or fall (S)	10-25	Winter annual in south. Best on good soil (not wet or alkaline). Good with ryegrass or small grains.
lespedeza	spring	15-40	Annual or perennial varieties. Tolerates poor and acid soil. Can be used as a forage.
medic	spring-fall	15-25	Most varieties annual, does not overwinter. Does not tolerate drought. Good with grasses in pasture.
red clover	spring/late summer	8-15	Some tolerance of acid & wet soil. Good warm season growth for hay or pasture.
soybean	spring	60-100	Does not overwinter. Good as a forage.
subterranean clover	spring/fall	10	Winter annual; reseeds itself. Tolerates poor or acid soil. Good with grasses.
sweet clover	spring/ summer	8-25	Annual (white) or biennial (yellow). Tolerates dry soil. Good with small grains for forage. Has toxic coumarin, avoid over-feeding to livestock.
vetch	spring/fall	20-60	Good with oats or rye but can become weedy. Can be used as a forage.
white, Ladino clover	spring/late summer	10	Tolerates dry soil. Good with grasses or interseeded in row crops.

Non-legumes

Crop	Planting	Seed lbs/acre*	Comments
barley	spring	60-100	Does not tolerate acid soil. Needs well-drained soil.
brome grass	spring	10-15	Good with alfalfa. Good as a forage.
buckwheat	spring/ summer	35-75	Does not overwinter. Tolerates poor or acid soil.
millet (pearl)	spring	25-30	Fast warm-season growth. Good forage.
oats	spring/fall	75-100	Tolerates acid soil. Good nurse crop for legumes.
rape, radish, mustard	summer	10-20	Fast growth in cool weather. Good catch crop. Good as a forage.

COMMON COVER, SMOTHER & GREEN MANURE CROPS
(continued)

Crop	Planting	Seed lbs/acre*	Comments
reed canary grass	spring/fall	15-25	Tolerates wet soil. Good growth in cool weather. Good as a forage.
rye (winter or grain rye)	late summer/ fall	75-100 200-300	Tolerates poor soil. Can control weeds by allelopathy. Good as a forage.
ryegrass	spring/fall	20-25	Best in good, well-drained soil. Good catch (annual) crop. Good as a forage.
ryegrass	spring	25-75	Tolerates wet soil. Good as a forage. (perennial)
sorghum sudan grass	late spring	20-35	Fast-growing. Does not tolerate wet soil. Good for silage or green chop.
wheatgrass	spring/ fall	4-10 (drill)	Tolerates dry & poor soil. Good growth in cool weather. Good as a forage.

*drill or broadcast

Mechanical Control

This final non-toxic weed control category includes many methods that do not involve chemicals or cover crops, but which kill or injure weeds by some kind of physical disturbance.

Cultivation, etc. Cultivation and similar methods control weeds by disturbing the surface soil and either uprooting them, cutting them in pieces or allowing their roots to dry out. Hilling and ridging operations also kill weeds by burying them. It usually takes two or three cultivations (or rotary hoeings) to kill most of the weeds early in the season, since new ones sprout or ones missed by earlier cultivation keep growing. Timing is very important, since the young weeds are easy to control before their roots go deep. Care must be taken not to prune or disturb crop roots. In-row weeds are more difficult to control than those between rows, but some methods do a good job. Shields can be used to protect young crop plants.

Cultivation has an additional benefit. Besides controlling weeds, it also loosens and aerates the soil. Tests have found yield increases of 10 to 20 bushels per acre for corn and two to five bushels for soybeans. Sometimes the yield benefit from cultivation comes more from soil aeration than weed control.

There are many kinds of cultivation machinery. Some are tricky to use effectively, and wet weather can cause havoc by preventing timely cultivation. The most commonly used cultivation equipment includes:

Rotary hoe — good for very young weeds. Not effective in wet or compact soil. Correct speed is important, seven to nine mph. Rotary hoeing should be done three to five days after planting and again when the row crop is two to three inches high. Hoeing when the crop is very small or going too deep can damage or thin the crop. Having a loose, crumbly soil works best, since weed seeds in the top two or three inches are the ones that germinate. Hoeing early in the morning and at fast speed loosens surface soil and allows the sun to dry and kill the weeds. If it rains after you rotary hoe, you probably will need to do it again.

Shovel and sweep cultivators, field cultivators — good for young weeds and to hill. Many farmers combine cultivation with side dressing nitrogen or interseeding a cover crop so as to save trips over the field.

Disk hiller — a nearly horizontal disk for weed control and hilling. Many biological farmers think the rotary hoe and disk hiller are the best-kept secrets for controlling weeds with no or minimal herbicides.

Spike-tooth harrow — good for very young weeds. Not effective in wet soil.

Rod, finger, rolling and spring-hoe weeders — narrow or vibrating rods can even get in-row weeds. As with rotary hoeing, timeliness may be important. These tools don't work well with lots of surface residues.

Plowing and disking — can kill deep-rooted or perennial weeds.

Burning. Using a hot propane flame will instantly kill or injure young weeds (especially under two inches tall). When a crop such as corn is at least about 10 to 12 inches tall, the rapid pass of the flame will not seriously harm it. A flame cultivation rig usually has two burners per row, one staggered ahead of the other, and aimed downward toward the row at an angle. Varying the travel speed is one way to control temperature. Burning is generally used for in-row weeds that cultivation cannot reach. It works especially well on grass weeds. One advantage is that crop roots are not disturbed. Burning works best on hot, dry days. Propane costs run about $3.00/acre. Burners can also be used to kill certain insect pests, such as potato beetles. Like any other tool, learning how to use a burner effectively is important.

Mowing. If weeds have gotten beyond the young stages where cultivation or burning are possible, mowing can be a way to weaken them or to reduce their seed production. Mowing is mainly feasible in places without ordinary cultivated crops, such as in barnyards, vacant lots, fallow fields, pastures, road sides or between fruit trees or vegetable beds. To be effective, mowing should be done before the weeds have started to set seed. Mowed weeds that have set seed can bring them to maturity even after they have been cut down.

Herbivores. Animals that eat plants (herbivores) can be an effective method of weed control under certain limited conditions. Some fruit or vegetable growers use "weeder geese," goats or sheep. Weed control scientists are trying to find insects or diseases that attack only weeds (not crops). They have had some success by importing foreign insects to control foreign species of weeds. A type of seed-eating weevil has been released in the midwestern United States to control musk thistle. Some farmers have gotten good control of Canada thistle by letting native butterfly caterpillars eat the leaves.

Hand-weeding. Few people would consider hand-weeding anything larger than a garden, but some fruit and vegetable growers find that hiring people to hand-pull or hoe weeds in high-value crops is cost-effective (and leaves no toxic residues). A few grain farmers use hand-weeding to eradicate certain very noxious weeds that have just begun to invade a field, before they become very common.

Herbicides as a Back-up

As a final note to this chapter, we need to emphasize that effective weed control generally requires a multi-pronged attack. Different weather conditions each year mean that weed problems may differ considerably. Different crops in a rotation have their own set of weeds. And different soil types and slopes on your farm may have an effect.

Most biological farmers try to improve their soil structure and get nutrients into balance as long-term control measures. They plan their crop rotations to keep weeds off-balance, and they plant cover or smother crops. But on a yearly basis, they plan on using mechanical control methods to control that season's weeds, usually cultivation and/or rotary hoeing.

If it appears that non-toxic methods will not give effective control, most biological farmers do not hesitate to use herbicides as a back-up or supplementary control measure. Most do not like using toxic chemicals, but herbicides are less damaging than insecticides and fungicides, and it is better than large yield reductions. Of course, the organic farmer cannot use herbicides at all.

Most biological farmers have devised ways of using less herbicide than the "full rate." The chemical company probably says on their label that their product is not guaranteed if used at less than recommended rates, but research and farmer experience have shown that effective control can be obtained down to half rate, if environmental conditions are favorable.

The effectiveness of most herbicides can be increased if a surfactant is mixed with them, either a synthetic surfactant or a natural one such as humic acid. This may allow you to get good control at one-half rate or less. Adding nitrogen fertilizer to the herbicide tank increases the effectiveness of some herbicides.

Another good way to reduce per acre herbicide use is to band-spray in the row and use cultivation or other mechanical methods between the rows. By using half-rate herbicide only in the row, the total per acre rate can be cut to one-fourth or less.

Finally, many biological farmers reduce herbicide use by only spot-spraying the worst areas of a field or by using a wick and wiper system to get a systemic herbicide onto weeds that are taller than the crop.

The Future

Researchers are developing some additional ecologically sound weed control methods, with an emphasis on biological control. A type of Mediterranean gall-mite is being studied which attacks bindweed and not crop plants. Certain bacteria produce enzymes that weaken the protective waxy covering of leaves, making them more susceptible to herbicides. Some fungi are being harnessed to kill weeds, although it is difficult to make spray preparations that have a long shelf life. However, a fungus-produced chemical, ALA (aminolevulinic acid) shows promise as a natural herbicide. When sprayed on plants, ALA causes plants to "overdose" on sunlight and self-destruct. Computerized field mapping and precision spraying are recently developed methods that can reduce the amounts of herbicides that are used.

Those are some pointers for environmentally sound, but profitable, weed control. Weeds can teach us valuable things. They will never be eliminated, and doing battle with them each year is a challenge.

20

PESTS, DISEASES & SOIL FERTILITY

Every agriculturist is concerned about crop pests as well as diseases. They annually cost producers billions of dollars, not only in lost production but also in fighting them — money spent for pesticides, fuel and labor — and the costs to society of environmental pollution and contamination of food by toxic residues.

Is there a way to avoid these costs and problems, but still protect crops from pests and diseases? Why and how do pests attack crops anyway? Are crop pests inevitable? The complete answers to these questions are not found in textbooks, nor are they expounded by agronomists, plant pathologists or agricultural entomologists. We need to look at evidence from many sources and piece together a picture, in jigsaw fashion. We still do not have all the pieces of the puzzle, but by combining scattered scientific research and the practical experience of biological farmers, a rather surprising picture emerges.

What is a Pest?

Over a million species of insects have been discovered by science. They eat all sorts of food, including dead carcasses, animal blood, wool, dung, other insects and wood. But a large number eat plants. Virtually all types of plants have insects that feed on them. When an insect eats something we don't want it to eat, we consider it a pest — whether wood-eating termites, blood-sucking mosquitoes or sap-sucking aphids.

The great majority of insect species — well over 95 percent — are not pest species; in fact, some are quite beneficial, since they provide useful products (honey, silk), pollinate plants, kill pest species, help clean the environment of wastes (dung, dead animals and wood) and provide food for many other wildlife species.

The relatively few species that are crop pests are often very abundant in numbers of individuals and cause devastating damage. Pests can cause damage ranging from minor yield reductions to death of the entire plant. Most damage is caused directly by the feeding of the insect, but some pests also carry virus and bacterial plant diseases. Further damage of agricultural products can occur when insects eat stored food, such as grain, meal, flour and fruit.

Pests Are Picky

Insects, including the pest species, are part of the earth's ecosystem, a complex network of interrelationships between the many species of plants, animals and microbes inhabiting the globe. We sometimes refer to this as the balance of nature. Plants manufacture food by photosynthesis, herbivorous animals (including insects) eat the plants, predators eat the herbivores, and scavengers clean up the wastes and recycle them back to the soil.

Casual observation may suggest that all plants are eventually attacked by insects, yet a hike through a meadow or a survey of a garden or field reveals an amazing fact: *not every plant is attacked by pests or diseases.* Some gardeners and farmers are so used to seeing their crops demolished by insects that they think pests are inevitable. This mind-set is even common among agronomists and entomologists, but a search of books and scientific research journals reveals occasional notes and studies tucked away that are pieces of our puzzle.

Sometimes only one plant in a row is not infested with aphids or leafhoppers. Sometimes a certain crop variety is resistant to a pest or disease. We need to discover *why.* Why do some plants resist pests, while others are susceptible? What can we do to increase crop resistance to pests and diseases? Some answers will emerge as we complete our jigsaw puzzle.

To return to the concept of an ecosystem, we might ask what is the place — or function — of plant pests and diseases in the balance of nature? An idea that has been popular among biologically oriented farmers and gardeners — an idea born of years of observation and experience growing crops — is that, in general, crop pests and diseases play the role of guardians, not wanton destroyers. That is, *they are there to selectively attack and destroy the weak and unfit,* plants that are "sick" and that would not produce good food for humans or animals. This idea may be ridiculed and dismissed by most professional agronomists and entomologists, yet once again, we can follow its thread here and there among the scientific literature.

So often, crop *quality* is given little or no consideration in modern agriculture. Instead, bulk — or yield — is the name of the game: *quantity* rather than quality. Perhaps the over-specialization of science and technology is partly to blame. The agronomist is only concerned with producing crops, not with what happens to them (who or what eats them). The nutritionist and doctor are interested in the diet and health of their patients, not how the food is grown. A misguided, profit-oriented market system must share part of the blame for failing to see the whole picture.

Insects and Their Food

What do insects require from their food, and how do they decide exactly what to eat? The answers are complex and hard to discover, since there are so many kinds of insects. For the most part, insects need the same basic nutrients as any other animal or human: carbohydrates, fats, proteins (or amino acids), vitamins, minerals and water. But the exact food materials the insect consumes to fulfill its nutrient requirements are another matter. As mentioned earlier, they may range from blood to dung to plant sap in various insects.

The exact food eaten may depend on the particular species of insect, its sex, its stage (larva or adult), whether it is moderately or very hungry, and environmental conditions (such as temperature, light, humidity, etc.). Insects "prefer" some foods over others. Sometimes the preferred foods give them faster growth or better reproduction. Some crop pests attack only one or a few kinds of plants, such as northern corn rootworm larvae, which only feed on corn. Other pests, such as grasshoppers, attack a wide variety of crops. How does an insect choose which plant it will eat?

Choosing Lunch

Exactly how an insect picks out what it will eat is not an easy matter to discover. There are many factors involved. Scientists have tried to pinpoint food selection factors mainly by doing laboratory experiments in which insects are given a choice between two substances. The plants or substances are then analyzed chemically to enumerate their components. But were the insects "happy" and "normal" in their confining laboratory cage? Perhaps they did not react as they would in nature. Of course, field experiments are sometimes done also, but you can see the difficulty in studying this sort of thing. Other scientists have approached the subject by measuring the nerve impulses of the insect's microscopic sense organs while stimulating them with different chemicals. It's not easy to discover how an insect chooses lunch.

The traditional view of how insects find and decide what to eat is that they use their senses of vision, smell and taste (especially the latter two). A few plant-feeding insects definitely use vision to recognize their food from a distance. One type of grasshopper has been found to be attracted by a pattern of vertical stripes, such as would be found in a bunch of grass. Some aphids are attracted by the color yellow. The reason is that they prefer more yellowish-green leaves (those that are either very young or old) rather than dark green leaves. Remember this fact; it is a piece of our puzzle.

Most insects are believed to recognize their food plants by smell — by a characteristic odor given off by the plant and sensed by the insect from a distance. If there is a breeze, the insect may simply follow the odor up-wind and come to the plant. When it is on the plant, the insect may smell it again to make sure it is suitable food. Then the insect will start to eat the plant, tasting it at the same time. If the plant is bad-tasting, the insect may stop at only a nibble and fly away. Both smell and taste involve the insect's sensing the chemical make-up of the plant.

There is more to the story. Research by one entomologist, which has been ignored by most insect physiologists, indicates that insects have an additional sense to help them locate food. They "tune in" to infrared frequencies given off by objects around them. They sense infrared radiation by special microscopic hairs on their antennae. Dr. Philip Callahan, former USDA entomologist in Gainesville, Florida, has found that odor molecules radiate ("broadcast") infrared wavelengths, and the exact frequency depends on their type, their concentration, the temperature, and how they vibrate. An insect's "odor" detection may involve more than the ordinary sense of smell. Even in the dark, a corn borer moth for example, can tell a corn plant from a lilac bush.

Again we ask, what exactly causes an insect to choose one plant over another to eat? Most entomologists believe that there is an interaction of two kinds of chemical substances produced by the plant: *attractants* and *deterrents*. Apparently, sometimes one or the other operates, and likely both play a role. For example, cabbage butterflies are attracted to plants in the cabbage family because they contain mustard oils. Substances called cucurbitacins attract certain beetles (including the cucumber beetle, also called the southern corn rootworm beetle) to the flowers of plants in the squash family (squashes, pumpkin, cucumber, melons).

Actually, a plant's odor consists of a blend of chemicals. Twenty-four substances have been measured in the air around corn plants, and 70 from cotton plants. Some insects may not respond to a plant unless all the components of the odor blend are present. Certain chemical by-products of plants are known to repel insects, to prevent their feeding or to cause poor growth and reproduction if they do feed. For example, Colorado potato beetles and potato leafhoppers are deterred by several alkaloids and saponins. High concentrations of attractants can act as repellents. In fact, some entomologists think what we call "food selection" by certain insects is really more the case of insects avoiding those plants that smell or taste bad.

Experiments have proved that insects definitely do better when they eat food from one type of plant than from another, and that complete nutrition is important. Grasshoppers actually lose weight when fed certain native prairie grasses; they do best on a diet of mixed species. When Colorado potato beetles are reared on young potato plants (with a high lecithin content), they later lay about 30 to 50 eggs at a time. If fed on older plants with a lower lecithin content, they only produce about eight to 20 eggs. And when fed on a different plant species (a close relative to potato), they lay no eggs. Aphids reproduce faster on young and rapid-growing *Euonymus* leaves than on mature or slow-growing leaves.

Plant Metabolism

Some entomologists have stated that whether or not pests attack a plant is connected with physiological (functional) changes in plant cells, yet botanists and agronomists tend to think that each species of plant always functions in the same way. If you dig into research papers and consult plant pathology books, you will find that plant metabolism and the metabolic products a plant produces can vary widely, both among individual plants of the same kind (corn plants in a row), from hour to hour and season to season, and in different varieties within the same species, not to mention between different species. For example, in a study of odors given off by cotton plants, greater amounts were detected in July and August, when plants were producing buds and fruit (bolls). This is the same time boll weevils are attracted to cotton. Variation in a corn plant's chemical composition, from day to night and throughout its development, affects the feeding behavior of European corn borer larvae.

Among the factors that can determine and alter how a plant functions are its genetic makeup (as in individual and variety differences), light intensity and duration (affects photosynthesis), temperature, water, other organisms (including pests, disease pathogens and beneficial organisms, such as nitrogen-fixing bacteria and mycorrhizae), and *soil conditions*. Important soil conditions that affect plant functioning include aeration (along with oxygen and carbon dioxide concentrations), water content (with extremes of waterlogging and drought) and *fertility* (the nutrients present and their balance). Washington State University plant pathologist R. James Cook has written that drought stress limits many normal functional processes of plants (*Phytopathology*, vol. 63, p. 455).

The Importance of Soil

In the past, some agronomists or other specialists had trouble believing that soil fertility — as long as the plant had adequate levels of necessary elements — could have any significant effect on the plant's metabolism, resistance to pests and diseases, or the nutritional value (quality) of the crop. Some people may still believe this, but many experiments over the last 50 years have shown otherwise.

Differences in soil condition and soil fertility can affect such plant characteristics as general growth rate and height, time of maturity and ripening, resistance to pests and diseases, total yield, nutritional value of food (protein, vitamins, mineral content, etc.) and keeping quality of food. Just to mention a few of the studies that prove this, three scientists from the University of Guelph, Ontario, Canada, compared nitrate and ammonium fertilizer sources of nitrogen for young corn plants (reported in *Agronomy Journal*, vol. 62, pp. 530-532). They found that there was no difference in overall yield, but the nitrate fertilizer caused the corn to take up about 14 percent more nitrogen than the ammonium fertilizer did; however, about 25 percent of it remained as inorganic nitrate nitrogen rather than being changed into protein. Also, the nitrate fertilizer resulted in more calcium and magnesium in the total plant (but less magnesium in the top part), while ammonium caused more phosphate and sulfate in the entire plant (but no significant difference in sulfate in the roots).

USDA researcher C. B. Shear states that "inadequate Ca [calcium] in fruits is a major cause of reduced quality and shortened storage life" (*Communications in Soil Science and Plant Analysis*, vol. 6, p. 233). He further lists several diseases in apples, cherries, plums, pears, avocados and mangos that are associated with low calcium. He also says that the relative balance of calcium to other nutrient elements is very important in growing quality crops (*ibid.*, p. 236).

A study in New York comparing two magnesium fertilizers, magnesium sulfate (epsom salts) and dolomitic lime, found that potatoes grown with magnesium sulfate were higher in quality (less discoloration, lower phenols, and higher in lipids, phospholipids and magnesium) (*American Potato Journal*, vol. 64, pp. 27-33). A study of nutrients in alfalfa and orchard grass found that the crops had a higher amino acid content when the potassium-to-nitrogen ratio in the fertilizer was 1:5 (*Crop Science*, vol. 7, pp. 599-605). Studies in which differently fertilized crops were fed to animals to test nutritional value have found that balanced soil fertility with abundant calcium produces superior food (for example, see *Agronomy Journal*, vol. 48, pp. 147-152; *Proceedings of the Soil Science Society of America*, vol. 6, pp. 252-258; vol. 7, pp. 322-330).

How a soil is fertilized can also affect its structure, and therefore affect such conditions as aeration, drainage, plant water uptake and root growth. In certain soils, the use of high ammonia fertilizers (especially anhydrous ammonia) can cause leaching of humus and hardened soil (*Plant and Soil*, vol. 29, p. 468; *Soil Science*, vol. 102, pp. 198-199). High levels of soil magnesium, whether from local subsoil or applied lime and fertilizers, can lead to tighter soil in some cases (certain types of clays) (*Russell's Soil Conditions and Plant Growth*, 11th edition, p. 939; *Australian Journal of Soil Research*, vol. 11, pp. 143-165; vol. 156, pp. 255-262).

Soil Life

Another aspect of the soil that many researchers do not consider is soil life — the multitude of soil organisms that directly or indirectly influence plant growth — and therefore crop quality and nutritional value. Normal fertile soil is teeming with an amazing array of plants and animals, most of them tiny or microscopic. They include bacteria, fungi, algae, protozoans, nematodes (roundworms), earthworms, mites and insects. Most of them are very beneficial to soil conditions and to plant growth and health. Bacteria and fungi are especially important.

Benefits derived from soil organisms include improved soil structure (aggregation of fine particles), better aeration and water intake, erosion resistance, drought resistance, organic matter recycling, increased humus content (which improves all of the above as well), better plant nutrition (increased amount and availability of certain elements), buffering against extremes of pH and salts, and protection of roots from certain insect pests, nematodes and diseases.

The soil is not just a medium to hold up the plant and supply a dozen or so elements for growth. The soil is a complex "nursery" that promotes optimal plant growth and health. Soil consists of chemical components (nutritive elements and ions), physical factors (structure, aeration, etc.) and biological organisms (which recycle organic matter and improve the chemical and physical factors). As we shall see, when the total soil environment is satisfactory, not only are crop growth and quality good, but *plant resistance to pests and diseases is also enhanced.*

Natural Resistance to Pests and Diseases

The fact that certain plants are resistant to insect pests and diseases has been scientifically proved many times, although the exact reasons for natural resistance are not always easy to discover. Scientists often break natural resistance down into several categories to study, but probably in nature more than one type of resistance operates at one time. We briefly mentioned some of them earlier.

First, there is a *genetic factor* to natural resistance to pests and diseases. Some species and some varieties within a species have genetically controlled mechanisms that allow them to better resist pest and disease attack. Resistant varieties of certain crops are available to the grower.

There is another sort of resistance, sometimes called *induced resistance*. This would include external factors, whether natural or applied by man, that increase whatever resistance mechanisms a plant inherits genetically. Sometimes even a crop variety that is genetically susceptible to a pest or disease can become more resistant than it would be otherwise. The improvements in plant metabolism due to better soil fertility and beneficial soil organisms would fall under this category.

Next, what are the actual mechanisms of plant resistance to pests and diseases? Some of them are fairly simple, and others are quite complex and still not well understood. One mechanism is *barriers*. The thick, waxy outer covering of some plants protects them from germinating pathogenic fungal spores. A dense mat of hairs on the leaves and stems of some plants protects them from small insects such as aphids, leafhoppers, flea beetles and young corn borer larvae. Scientists are trying to breed hairiness into varieties of potato, alfalfa and other crops.

Another mechanism of resistance is *repellent chemicals*. Some plants produce strong-smelling odors or toxic chemicals that keep pests from even starting to attack. Examples are turpentine and resins found in conifer trees, mint oils, citrus oils, and garlic and onion oils.

Special plant chemicals called *phytoalexins* can kill or inhibit invading fungi, bacteria, viruses and nematodes. Many plants can produce them, including some crops. Corn produces two phytoalexins that give resistance to corn borer larvae, gibberella stalk and ear rot, and the northern corn leaf blight fungus.

Certain proteins and amino acids (that are not part of proteins) appear to give natural resistance to some plants. Even administering synthetic amino acids to diseased plants can sometimes control the disease. Some amino acids (phenylalanine and tryptophan) are used by plants to make phenolic substances, which are toxic to fungal pathogens. In potatoes, an imbalance of amino acids caused by high potassium fertilization was found to increase late blight disease in one study (*Annual Review of Phytopathology*, vol. 4, pp. 349-368).

Still other plant chemicals of various types can play a role in natural resistance. Some taste bad to insects or cause them to stop eating and starve. Others reduce the insects' fertility or egg-laying ability. Some inhibit the growth of fungal or bacterial pathogens. These miscellaneous protective chemicals include tannins, protocatechuic and chlorogenic acids, coumarin derivatives, protease inhibitors and polyacetylenes. Some are only produced by the plant's cells when the plant tissue is injured or attacked by pests or disease pathogens.

A similar mechanism is involved whereby chemical odors, produced by a plant only when it is injured by a pest, then attract parasites of that pest. USDA researchers found that corn seedling plants release certain kinds of terpenoids only after several hours of feeding by armyworm caterpillars. These odors are attractive to parasitic wasps which lay eggs in the caterpillars that hatch into voracious wasp larvae, which later kill the armyworms. The wasps are not much attracted to odors of artificially damaged corn plants or to odors of caterpillar oral secretions or feces (*Science*, vol. 250, pp. 1251-1253).

Another kind of resistance is overall plant strength and hardiness. A vigorously growing plant can often out-grow moderate pest or disease damage. Also, a strong stalk can withstand certain stalk-boring pests, such as corn borers, rice stem borer, wheat stem sawfly and the hessian fly.

If we review the several types of natural plant resistance to pests and diseases just covered, we note that most of them involve plant functioning on the cellular, biochemical level. The plant's physiology, or functioning, determines to a large extent whether of not it will resist attack. Many researchers have commented on this, some even noting cases of "temporary" resistance because of changes in plant functions (for example, *Journal of Agricultural Research*, vol 73, pp. 33-43; *Annals of the Entomological Society of America*, vol. 49, pp. 552-558; *Journal of Insect Physiology*, vol. 18, pp. 423-437; *Hilgardia*, vol. 32, pp. 501-537; *Canadian Journal of Plant Science*, vol. 44, pp. 451-457).

Stress Versus Health

One aspect of natural resistance that is often overlooked by agronomists and even plant pathologists is the relation of stress to the health of the plant — and the importance of plant health to natural resistance. The final link — the relation of soil conditions and fertility to plant health — is similarly usually not considered.

A good generalization is that *pests and diseases attack stressed plants.* This is just a more scientific way of stating what we mentioned earlier, that pests and diseases are there to eliminate the sick and unfit plants. There are many scientific studies, plus casual observations, that confirm this. For example, consider stress caused by either too much or too little soil water (waterlogging or drought). In a review article on the subject, R. D. Durbin notes that these stresses "clearly constitute a major limiting factor for plant growth and productivity," and that "in some cases they reach such a magnitude that they cause abiotic diseases" (*Water Deficits and Plant Growth*, edited by T. T. Kozlowski, vol. 5, pp. 101-117). Abiotic diseases are those not caused by pathogens (germs); they can have such symptoms as growth and yield reduction, wilting and yellowing of leaves, dropping of flowers or fruit, fruit rot and root death. Imbalances of internal plant hormones and abnormal cell changes are often involved, along with toxic levels of plant metabolic by-products. Under water stress, soil conditions become unfavorable, including low oxygen/high carbon dioxide, high salts, pH changes, toxic levels of nitrites and certain trace elements, and harmful (anaerobic) soil microbes. Dr. Durbin states that a stressed plant "is more easily damaged by other stresses, diseases, or insects, and these, rather than the water stress, are blamed for death of the plant." That is a significant statement.

This realization by plant pathologist Durbin points out the common problem of conventional agriculture. For the last 40 or 50 years at least, the usual approach of crop specialists has been to fight the *symptoms* (crop pests and diseases) with toxic chemicals (insecticides, fungicides, etc.) rather than discovering and correcting the real underlying causes (stresses due to wrong soil conditions). They simply fail to see the connection between soil conditions and crop pests and diseases.

Sir Albert Howard observed that crops stressed by waterlogged and poorly aerated soil were more seriously attacked by pests and disease. He believed that only plants that became physiologically abnormal were attacked. He noted that an injured root system preceded disease or insect attack (*Annals of Applied Biology*, vol. 7, pp. 373-389).

We could cite other stress-producing factors that can harm plants, such as cold or hot temperatures, and toxic effects of certain fertilizers (anhydrous ammonia, high-salt fertilizers) and chemicals (herbicides, pesticides). High salts increase the growth of some pathogenic fungi, including

Fusarium, Phytophthora, Sclerotinia and *Rhizoctonia* (*Phytopathology*, vol. 63, pp. 453).

Basically, it boils down to this: stresses lead to an unhealthy plant, and an unhealthy plant is more susceptible to invading pests and diseases. At first, a plant may not appear sick; it may have a "subclinical" ailment, such as upset internal metabolism. Perhaps certain enzymes or hormones are crippled. This means that its normal growth, and production of food and protective substances (phytoalexins, repellents, etc.) will be hampered or stopped. Essentially, the plant loses its natural immunity to pests or diseases, which would be present in a vigorous, healthy plant.

For example, several plant odors are produced by radish, turnip and cauliflower plants when tissue is injured that are *not* present when plants are normal, and these odors may attract pests. Aphids have been found in larger numbers on sugar beet plants already infected with yellows virus disease than on healthy plants.

Some stresses, such as floods, drought and other weather effects are beyond the farmer's control, but soil condition is one aspect which the farmer can and should control to minimize or prevent pest and disease problems. A balanced, healthy soil will help prevent the attack of pests and diseases. A "healthy" soil means one that is close to ideal in all three areas: chemical (nutrients), physical (structure) and biological (organisms). The ideal conditions for one crop may not be perfect for another, but the usual field crops grown in the U. S. have some common denominators. Generally, soil that is balanced and adequate to high in nutrients, well-supplied with humus (not just "organic matter," but decomposed humus), with a good population of beneficial soil organisms and few or no toxins, will produce crops that are high-yielding, high quality, healthy and pest- and disease-resistant.

Can Health be Measured?

The health of a growing crop is vital for not only yield and quality, but also for preventing pest and disease attack. Is there any way to measure and monitor crop condition *before* disaster strikes? If you had unlimited money and plenty of time, you could do complex tissue tests and find out the exact amounts of various amino acids, enzymes and a host of other indicators of plant metabolism. Even then, such tests would be hard to interpret and may not be of any value. Waiting a week or more for the results may be too late.

One quick method of monitoring plant health before full-blown disease strikes is to measure the infrared reflectance of the leaves. Stressed or sick plants reflect less infrared light than healthy ones, due to changes in leaf cell structure. Plants can either be photographed with infrared-sensitive film or instant measurements can be made with an infrared "gun."

In general, the health of the plant is a reflection of its overall functioning. This can be partly told by its photosynthesis (sugar production) and translocation (movement of sugar from leaves to other parts of the plant). Again, there are expensive, time-consuming scientific tests for such things, but they are impractical for the grower.

There is a simple, inexpensive way to monitor crop health — daily, or even hourly if you wish. This is by using a scientific instrument called a refractometer (see Chapter 8). Portable (hand-held) refractometers have been used for years by growers and technicians to measure the sugar content of crops for canning and beverage making (wine, beer), among other uses. Actually, a refractometer measures the percent of all soluble solids in a liquid. In plant sap, this will include mostly sugars (primarily sucrose, glucose and fructose), but also some salts, organic acids, amino acids and proteins. The refractometer measures soluble solids in units called Brix, which is equivalent to percent.

Studies with corn have found that during early growth, non-sugar soluble solids are fairly high (up to 50 percent) in the stalk, but after tasseling they decrease to a minority (about 25 to 35 percent), with sugars predominating (*Crop Science*, vol. 10, pp. 625-626). Percent sugars tend to be lower in corn roots and leaves and higher in the stalk (after tasseling), while within the stalk, sugar readings are highest in the middle half and lower near the top and bottom (*Canadian Journal of Plant Science*, vol. 44, pp. 451-457; vol. 52, pp. 363-368; *Plant Physiology*, vol. 5, pp. 555-564). But stalk sugar content also changes during maturity. Sugars accumulate in the stalk up until the kernels begin filling; then there is a decrease as the stalk sugar is translocated into the ear and changed into starch (*Crop Science*, vol. 9, pp. 831-834; *Agronomy Journal*, vol. 44, pp. 610-614). Sugar levels in leaves and the upper plant are also higher at peak periods of photosynthesis, during warm sunny weather.

There is, however, quite a bit of variation among different varieties of corn in the percent sugars accumulated in the stalk. There is an increase in protein content of the grain as sugar production increases (proteins are made mainly from sugar, plus nitrogen and sulfur). Grain yield is also correlated with sugar production (*Crop Science*, vol. 9, pp. 831-834; vol. 24, pp. 913-915).

Studies have been made to check the accuracy of refractometer readings compared to laboratory chemical analysis for sugars. They found a good correlation between Brix readings and the lab results, and concluded that for sugars above 1 percent and for all but very exacting studies, hand refractometers give reliable measurements of plant sugar content (*Crop Science*, vol. 10, pp. 625-626; vol. 24, pp. 913-915; vol. 28, pp. 861-863).

Little scientific research has been done to see if soil fertility is related to plant sugar content, although the experience of many biological farmers indicates that balanced fertility and high populations of soil organisms do help produce high-sugar crops, with associated high nutritional value for animals. Some Wisconsin dairy farmers report measuring the same milk production after switching to biologically grown feed as with conventionally grown feed, but *the cows consumed 10 to 15 percent less dry matter*. Biologically grown feed obviously has high nutritional value. In one scientific study, potassium applied as potassium chloride (muriate of potash) significantly decreased corn sugar content. This was caused by the chloride rather than the potassium, since ammonium chloride also caused a sugar reduction (*Agronomy Journal*, vol. 59, pp. 332-334). In another study, Kentucky bluegrass fertilized with two synthetic nitrogen fertilizers (ammonium sulfate and ammonium nitrate) had lower levels of sugars than control plants, while calcium nitrate fertilizer raised sugars slightly (*Phytopathology*, vol. 67, pp. 1239-1244). Fertilizers can influence the nutrient content of plants.

Sugar, Pests and Diseases

There has been some scientific notice of the relation of plant sugar content to the attack of pests and diseases. One study has found that northern corn leaf blight incidence is higher when leaf sugar content is lower (*Plant Physiology*, vol. 24, pp. 247-252). Several studies have found a strong correlation between the stalk sugar content of corn late in the season and resistance to stalk rot and lodging. Stalks that remain green and alive are very resistant to the invasion of stalk rot fungi. The reason appears to be that living stalk pith tissue is resistant, while dried out and dying tissue is readily attacked (*Phytopathology*, vol. 51, pp. 376-382; vol. 56, pp. 26-35; *Canadian Journal of Plant Science*, vol. 44, pp. 451-457). This supports the general principle that vigorous, healthy plants can resist diseases and that stresses lead to pest or disease attack because the plant's normal functions are upset. A steady, adequate level of sugar is believed to be necessary to maintain vigor; in fact, stalk rot seldom attacks corn plants that have had their ears removed (they have high stalk sugar because there is no ear to receive it) (*Canadian Journal of Plant Science*, vol. 44, pp. 451-457).

A number of other plant diseases are connected with low sugar levels, including wheat stem rust, onion pink root, storage rot of sugar beets,

Fusarium root rot in alfalfa, damping-off in tomatoes, melting-out disease of Kentucky bluegrass and leaf spot of finger millet. Plants with high sugar levels are resistant, and in some cases, spraying plants with a sugar solution controls diseases. Some other plant diseases appear to be worse in high-sugar plants, such as root rot of cotton. These studies have been summarized in *Physiology of Disease Resistance in Plants*, by P. Vidhyasekaran, vol. II, pp. 39-49, 71-72.

The reason high sugars often control disease pathogens can vary. Sometimes disease fungi cannot grow well with high sugar concentrations (over 10 percent). In other cases, sugars can inhibit fungal production of enzymes needed to break down the cell walls of the host plant. Another reason is that sugars are used by plants to produce phenolic substances that are toxic to fungi.

The role of sugar in insect attack of plants is fascinating. Based on research done on various insect and plant species, apparently insects like moderate amounts of plant sugars and are attracted to plants containing them. High concentrations of sugars are avoided by leafhoppers, grasshoppers and the European corn borer (*Annual Review of Entomology*, vol. 5, pp. 202-203). Exceptions include flower-feeding flies, bees and butterflies, which respond positively to high sugars.

Alfalfa was found to be resistant to pea aphids when its stem tissues had a more acid pH and higher levels of sugars (pentoses) and pectic substances (larger carbohydrate molecules formed from linked sugars) (*Journal of Agricultural Research*, vol. 73, pp. 33-34).

A possible reason some insects avoid high-sugar plants comes from research by G. Fraenkel. Some sugars and sugar-alcohol combinations (glucoside and mannoside) interfere with normal utilization of other sugars, and so are toxic to insects (mealworms). The inhibitory sugars are found mainly combined with other molecules in plants, but if digested by insects and in the presence of the sugar glucose, their toxic effects occur (*Journal of Cellular and Comparative Physiology*, vol. 45, pp. 393-408; *Journal of Experimental Zoology*, vol. 126, pp. 177-204).

More Than Sugar

Some growers may get the impression that sugar content and refractometer readings are all there is to preventing pest and disease problems. Some believe that if refractometer readings are above a certain number, say 12 Brix, that a crop will never have pests, but it is not always that simple. True, refractometer (sugar) readings are a good indicator of a plant's state of health and vigor, which are correlated with pest and disease resistance. However, refractometer readings will vary considerably among different plants in a field, in different parts of the plant, at different times of day and under different weather conditions. Anything that reduces photosynthesis will lower sugar levels, such as cool or cloudy weather and low light conditions. Normally, peak photosynthesis occurs in the early afternoon. The sugar produced in the leaves is then translocated to other parts of the plant and used for metabolism during the night. Sugar levels may reach a peak in the leaves in mid-afternoon, but not peak in the stem until evening. Therefore, when you take refractometer readings, be sure to do it at the same time of day, under similar weather conditions and from the same part of the plant each time. Something as small as whether you squeeze sap from only the corn leaf proper or whether you also include the midrib can make a difference of several percentage points in Brix readings.

Some abnormal or diseased conditions of the plant can cause high sugar readings, even though the plant is not healthy. A phosphorus or a potassium deficiency is known to cause sugar accumulation in leaves. Under stress or diseases where the plant's food-transfer vessels (the phloem) are plugged, sugar will accumulate in the upper parts of the plant.

Some pests appear to be attracted to the sugar content of plants, or are stimulated to feed in laboratory tests when given sugar, but more than this can be involved. Colorado potato beetles, European corn borer larvae and alfalfa weevils are reported to be attracted by plant sugars, but also by certain amino acids, lipids and mineral salts (*Entomologia Experimentalis et Applicata*, vol. 12, pp. 777-788; *Annals of the Entomological Society of America*, vol. 49, pp. 552-558).

The European corn borer moth is said to choose the largest and most vigorous plants to lay eggs on. But is the largest, darkest green corn plant really the healthiest? Many experiments have proved that plants given too much nitrogen fertilizer, although they may grow faster and look greener, are actually physiologically unbalanced and more susceptible to pest and disease attack than plants grown with balanced fertility (see for example, *Proceedings of the American Society for Horticultural Science*, vol. 63, pp. 304-308). Studies have found that higher levels of nitrogen fertilization and higher plant populations result in greater survival of European corn borer larvae (*Crop Science*, vol. 5, pp. 261-263; *Annals of the Entomological Society of America*, vol. 59, pp. 633-638).

High nitrogen fertilization is known to significantly increase the crop's free amino acid (non-protein nitrogen) content. In sap-sucking insects such as aphids, the relative sugar versus nitrogen (especially amino acid) content of the plant plays an important role in resistance to insect attack.

In order to get a balanced diet, aphids need a certain amount of amino acids, and plant sap has a relatively low level of amino acids compared to sugars. In a laboratory preference test, aphids preferred sugar (sucrose) over plain water, but preferred sugar plus six amino acids much more than just sugar and just amino acids (*Entomologia Experimentalis et Applicata*, vol. 7, pp. 315-328).

Earlier we mentioned that aphids sometimes prefer young or older leaves of a plant (especially young), not "middle-aged" leaves. The activities of leaf tissues and their chemical content varies with age. Plant sap used by aphids is richer in nitrogen compounds in young and older leaves than in middle-aged leaves (*Journal of Experimental Biology*, vol. 35, pp. 78-84). Middle-aged leaves have the highest photosynthesis rate and highest sugar content. Also, deterrent chemicals appear to be the most abundant in middle-aged leaves (*Annals of Applied Biology*, vol. 38, pp. 61).

In tests of corn earworm attack, more resistant corn varieties had higher sugar levels and lower amino acids (*Journal of Economic Entomology*, vol. 59, pp. 1062-1064; *Annals of the Entomological Society of America*, vol. 58, pp. 401-402).

Balanced Soil Fertility

Total soil fertility is important. In a study of varieties of alfalfa, either resistant or susceptible to aphids and grown in different nutrient combinations, unbalanced nutrients altered the crop's resistance. The susceptible variety remained susceptible, but the resistant variety became less resistant with low levels of calcium or potassium, or with high levels of magnesium or nitrogen. Low levels of phosphorus increased resistance, while varying sulfur had no effect (*Journal of Economic Entomology*, vol. 63, pp. 938-940).

USDA researchers K. T. Leath and R. H. Ratcliffe state that "balanced nutrition is probably more critical and has greater impact on disease development than does the actual amount of any single element" (*Forage Fertilization*, edited by D. A. Mays, pp. 482-483).

In various studies, sweet clover weevils were less damaging to new clover stands when soil fertility was high and balanced (*Journal of Economic Entomology*, vol. 47, pp. 117-122); nutrient-deficient peas were more heavily damaged by pea aphids than plants grown in high, balanced fertility (*Journal of Economic Entomology*, vol. 44, pp. 1010-1012; vol. 47, pp. 113-116); and New Zealand spinach grown in soil deficient in nitrogen and calcium was attacked by greenhouse thrips (*Journal of Economic Entomology*, vol. 39, pp. 8-11).

Quite a number of articles have been published showing the effectiveness of fertile, balanced soil in controlling crop diseases; many of them are summarized in the books, *Management of Diseases with Macro- and Microelements*, edited by A. W. Engelhard, and *Physiology of Disease Resistance in Plants*, by P. Vidhyasekaran, vol. II, pp. 65-74.

Unfortunately, only a relatively small amount of research is being done along the lines of pest and disease prevention by management of soil conditions and fertility. Few scientists can see the big picture and the interrelated ways the soil-plant system functions. Fortunately, many biological farmers are doing their own pioneering testing and experimenting and proving for themselves the validity and effectiveness of these concepts.

Summary

- Pests and diseases function in nature to eliminate the sick and unfit.
- Not every plant is attacked by pests or diseases.
- Vigorous, healthy plants have natural defenses against pests and diseases.
- Various stresses can break down the plant's immunity to pests and diseases. Stresses can come from (1) the weather (such as wrong temperatures, lack of sunshine and storms) or (2) from poor soil conditions (including too much or too little water, poor aeration, high salts and out-of-balance nutrients).
- The best long-range approach to crop pests and diseases is not to kill the attacking pests or pathogens, but to *prevent* the problems by correcting wrong soil conditions.

21

WEAK ENERGIES

Earlier we looked at some "fine tuning" methods and products (Chapter 16) that may help improve crop performance as a secondary measure to the basic principles of crop rotation, organic matter recycling, soil balance and promoting beneficial soil organisms. In this chapter, we consider some natural sources of plant-boosting energy that some growers have used successfully. Let me emphasize that I consider these approaches *secondary* to following the basic biological principles of agriculture that this book emphasizes. Do the basics first, and don't try to solve all your problems with a "magic bullet."

There is a whole world of energy that surrounds us and that influences plant, animal and human growth and well-being. We literally live in a sea of energy. Few agronomists and soil scientists take these natural energies into account when they design experiments, yet plant physiologists and physicists have documented that they can affect living things. One example that has been in the news lately is the possible harmful effects of strong electromagnetic fields.

The Electromagnetic Spectrum

Besides the obvious light energy we receive from the sun, there are also invisible ultraviolet and infrared wavelengths, and natural and man-made radio and microwaves, X-rays and gamma rays. These are called electromagnetic radiation, and they occur in different wavelengths or frequencies.

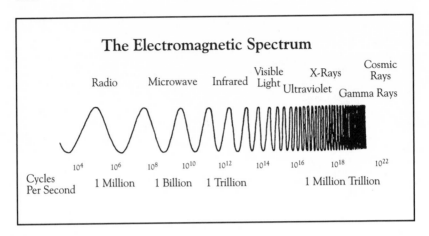

The Electromagnetic Spectrum

Everything in the universe — stars, rocks, plants and ourselves — radiates electromagnetic waves at a frequency depending on its temperature and caused by the vibrations of the atoms or molecules of which the object is composed. Very hot objects radiate infrared, visible light, ultraviolet, X-rays or gamma rays. Cooler objects give off infrared, microwaves or radio waves.

Besides these energies, there is electrical energy: static electricity, lightning, d.c. and a.c. Electrical charges possess a force field around them. Fluctuating electric charges can be used to generate electromagnetic radiation; this is how a radio station broadcasts.

The earth has a magnetic field. All objects on the earth are subjected to it. A change or movement in an electrical charge generates a magnetic field, and a change in a magnetic field generates an electrical field. Charged particles (electrons, protons, ions) in motion are affected by a magnetic field, and charged particles themselves can create an electric field that affects living things. It is known that an excess of positive ions (common in modern buildings and certain dry south winds) at first causes people to feel energetic, but after a while they "overload" the body's metabolic processes and hormone system, leading to anxiety, migraine headaches, arthritis flare-up, asthma, high blood pressure, higher accident and suicide rates, etc. Negative ions (common in mountains, forests and near waterfalls) or a "natural" balance of five positive to four negative ions, cause a sense of calmness and tranquility, sound sleep, and can help heal burns and asthma. Plants may grow unusually large with positive ions.

Cosmic rays (nuclei of atoms stripped of their electrons) bombard the earth from outer space, along with other charged particles: protons and electrons from the sun (the solar wind). The beautiful aurora borealis (northern lights) results when these particles interact with the earth's magnetic field and atmosphere. The sun also has a strong magnetic field which is variable; it reaches and influences the earth's magnetic field. The gravitational variations of the orbiting moon can affect the growth and behavior of plants and animals.

Bio-Energy Effects

Living cells and organisms can be affected by these weak natural energies in various ways, or even produce them. Living cells (and the larger tissues or organs that cells together make up) generate weak electrical and magnetic fields. Ions (electrically charged molecules or atoms) can be held or "pumped" to one side of a cell membrane, creating an electrical charge. This is the origin of the electrical impulses in animal and human muscle and nerve cells. Plant cells also generate electrical currents. The inside of pea root cells is about -110 millivolts (negative 110 mv) compared to the outside, while oat root cells are about -84 millivolts according to one study. Trees have daily and seasonal cycles of electrical activity, perhaps related to the flow of sap. Scientists have found that the electrical field of roots aids them in normal growth and nutrient absorption activities. Soaking roots in salt solutions (including potassium chloride, the fertilizer called muriate of potash) quickly lowers the root's electrical charge, interfering with nutrient uptake and tip growth. The thread-like cells of a fungus also generate electrical currents that enter at the tips and exit farther back.

Weak external electrical currents and magnetic fields are known to affect cell activity. The right kind of current can speed up cell functions, while microwaves have been found to be absorbed by DNA (the material that makes up genes), with possible damage and cancer initiation. Sending low-voltage pulsating electrical currents into the body through electrodes in the skin has been able to control the pain of neuralgia, headache, childbirth and arthritis, as well as helping circulatory problems, speeding healing of bones and burns, and improving memory. Even low-strength electrical fields can alter cellular enzyme functions and change the shapes of other large molecules, including some found in cell membranes and involved in vital functions such as DNA and protein synthesis (*Science*, vol. 236, pp. 1465-1469; vol. 247, pp. 459-462). Weak electromagnetic fields can alter the flow of calcium ions in brain cells, cause animals' body temperatures to rise, increase the number of white blood cells (involved in immunity) and slow down learning ability. It is no wonder that many people are concerned about the effects of living close to power lines and using electronic gadgets.

Lightning and thunderstorms generate electrical fields and radio waves (VLF, very low frequency). Scientists have recorded sudden changes in electrical fields of tree trunks during storms (*Science*, vol. 124, pp. 1204-1205). An external weak electric current has been found to stimulate root absorption of cations (calcium, potassium, sodium)(*Soil Science*, vol. 71, pp. 371-375; vol. 75, pp. 443-448).

Way back in the early years of the 20th century, experimenters in England treated seeds with weak electric current for several hours (seeds were soaked in a weak brine to carry electricity). Results included an increase of six to nine bushels/acre of oats, six to 12 bushels/acre of wheat, and a 21 to 50 percent increase of barley. Other more recent experiments have been performed, and an electronic seed treating machine has been developed, but the impracticality of such treatment seems to have held back its acceptance (*Acres U.S.A.*, Nov. 1975, pp. 12-13; Nov. 1976, pp. 18-19). Up to 100 percent faster germination and increased yields of corn, soybeans and vegetables, with less fertilizer required, were some of the reported results.

Living things are also affected by external magnetic fields. Tiny iron particles have been found in the cells of certain bacteria and algae, and in the heads of fish, sea turtles, birds, rodents, monkeys and humans. They are probably involved in the migrations of birds and sea turtles. Seeds placed in a moderate magnetic field for 10 days germinate faster than untreated seeds, and the seedling plants grow faster (for strong fields, a shorter exposure time is sufficient). In one test, magnetically treated barley roots grew 34.7 mm in four days, compared to 29.35 mm for untreated plants, and the stems of treated plants grew 21.1 mm long, while controls only grew 15.5 mm. Soybeans did even better: treated roots grew 22 mm, compared to only six mm for untreated. Results are better if the long axis of the seed is parallel to magnetic lines of force (*Hort. Science*, vol. 2, pp. 152-153; *Canadian Journal of Plant Science*, vol. 43, pp. 513-518; *The Science Teacher*, May 1973, pp. 57-60; *Soviet Plant Physiology*, vol. 22, pp. 84-89). Magnetic seed-treating machines have been developed and have been used in Canada for commercial treatment, but never did catch on in the United States (*Acres U.S.A.*, Apr. 1972, pp. 16-17; Aug. 1976, pp. 26; Aug. 1989, pp. 16-17). From 5 to 10 percent yield increases were reported for barley and 5 to 17 percent for corn.

Some crops mature earlier and produce higher yields when magnetically treated. Their cells appear to function more efficiently than untreated cells, producing more sugar but using less energy for metabolism. For example, tomatoes ripened faster in a magnetic field; they had a lower acidity and a higher sugar and carotene content (*Economic Botany*, vol. 122, pp. 124-134; *Nature*, vol. 199, pp. 91-92).

It is known that some plants' roots will mainly grow out of the seed in a north-south plane (winter wheat, some spring wheat, wild oats, some weeds), while others do not (barley, flax, corn, sunflower). If crops whose roots do grow north-south are planted in rows running east-west, they will more effectively use the between-row soil nutrients and water (*Crops & Soils*, vol. 20, no. 8, pp. 8-10).

Some researchers claim that the south pole of a magnet increases growth, while the north pole decreases it (they define the south pole as the pole that turns toward the earth's north pole, since opposite poles attract; however, most scientists designate the poles oppositely). In one experiment, crops grown from seed treated with the south pole of a magnet had better growth, yield, protein, sugars and oils than those treated with the north pole. Laboratory rats living in south-pole magnetism were larger and healthier than those living in north-pole magnetism, although they died sooner (A. R. Davis & W. C. Rawls, Jr., *Magnetism and Its Effects on the Living System*, 1974, pp. 31, 37-39).

Different molecules or substances have a "natural frequency" at which they vibrate (resonate), and they absorb electromagnetic waves more readily at that frequency than others (*Science News*, vol. 125, pp. 248). Some people speak of cells or whole plant or animal bodies as having a certain frequency. In the 1920s, the Russian engineer, Georges Lakhovsky, theorized that the nucleus of a cell is like an electrical oscillating circuit, emitting and receiving electromagnetic vibrations at a certain frequency if the cell were healthy. If the cell were unhealthy or dying, there was a lower frequency. He built machines which emitted short wavelength energy and was able to cure plant cancers, make plants and animals grow faster, and cure hospitalized patients of cancer, radium burns, goiter and other "incurable" ailments. The best results were obtained with a machine called a multiple wave oscillator, which allowed every cell in the body to vibrate at its own best frequency (P. Tompkins & C. Bird, *The Secret Life of Plants*, 1973, pp. 199-201).

Some agricultural workers have developed devices that are said to operate on these principles. Radionics machines are said to allow a person to detect and influence the energy fields of plants, animals, pests or even soil fertilizers (*Acres U.S.A.*, Nov. 1991, pp. 20-21). Devices called "towers" or "pipes" are said to "broadcast" beneficial frequencies or to collect and transmit "cosmic" energy to crops or animals.

Some people believe that "cosmic" energy also comes from the sun, a weak electromagnetic energy that bathes the earth and helps plants grow. Dr. Philip Callahan has researched the subject and believes that this energy is more readily channeled into plants by soils with high paramagnetic properties (a paramagnetic substance is a non-metallic material that is weakly attracted to a strong magnet). Clays and volcanic soils are highly paramagnetic and very fertile. Humus, oxygen and nitrogen also may contribute to plant growth in that way. By increasing your soil's aeration and humus level, you may be unwittingly magnetically treating your crop. Muriate of potash is said to lower the paramagnetic effects of soil (*Acres U.S.A.*, Nov. 1992, p. 5).

Another unusual application of weak energies to agriculture involves the use of certain sound frequencies to stimulate plant growth. Sound waves are a mechanical movement of air molecules (or other matter), so they are not electromagnetic in nature. Nevertheless, sound can have an effect on living organisms. It is well known that certain types of music (especially rock-and-roll) have harmful effects on our bodies, affecting our heartbeat, brain waves and glandular secretions, while most classical and jazz music has beneficial effects. The wise dairy farmer will tune the radio in his barn to soothing music. An Indiana University test found that cows increased their milk production 5.5 percent listening to symphonies and lowered it 6 percent when subjected to rock music (they didn't even want to enter the barn).

Sound also affects plant growth. Loud rock-and-roll or unmusical sounds can cause erratic or slow growth and low yield (the plants even use water three times faster), while pleasant music can increase growth (compared to untreated controls). However, loud, pure tones can cause seeds to germinate faster, according to research at the University of North Carolina, the University of Ottawa, Canada, and the USDA Beltsville, Maryland, research station (*Canadian Journal of Botany*, vol. 46, pp. 1151-1158).

In a practical application of the use of sound, Minnesota plant breeder Dan Carlson has developed a method of combining sound with a special foliar spray, often producing exceptionally large vegetables and crop yields 100 to 200 percent higher than untreated plants. Carlson's system uses cassette tapes of oscillating high frequency sound along with canary-like tweets or actual music. The sounds are played in a field through loudspeakers (which can be mounted on a tractor), and about 15 minutes later the crop is sprayed with the foliar spray, which contains trace minerals and natural ingredients. Carlson believes the sound opens up the plants' leaf pores (stomata), allowing them to absorb more spray. Tests with radioactive tracers have found over seven times more spray absorbed than by controls (*Acres U.S.A.*, Aug. 1994, pp. 1, 6).

Those are some of the ways the largely unknown weak energies can be harnessed to improve crop growth. They are interesting to contemplate and may play a larger role in the agriculture of the future.

22

ORGANIC, OR NOT?

One question that many biological farmers are considering these days is whether they should become "officially" organic; that is, become certified by an organic organization. Although it has survived a rocky — often despised — past, organic farming has finally begun to be recognized as a viable brand of agriculture. Sales of organically produced foods have skyrocketed over the last decade, largely because of the public's concern about toxic and potentially carcinogenic chemical residues in their food though another reason is the better taste and nutritional value of organic food. In fact, the demand for clean and organically grown food is now so great that growers have trouble keeping up, and market prices have increased to very profitable levels. Organic sales in the United States are now about two billion annually.

What is Organic?

By itself, the word "organic" means coming from or related to life. As a form of agriculture, organic implies using "natural" methods and materials. Actually, organic agriculture is near one end of a continuous spectrum, ranging from chemical-intensive "factory" farms at one extreme, to "low-input" conventional methods, to what we call "biological" (or others call ecological) agriculture, to organic at the other extreme. In the biological system that I have been explaining in this book, we use a combination of "natural" and man-made materials and, if necessary, chemicals can be used. Thus, biological agriculture is more "middle-of-the-road," but many biological farmers are nearly organic. All they may have to do is drop a couple of fertilizer materials and use additional organic nitrogen sources.

There are many kinds of farmers who call themselves organic. Some may use small amounts of synthetic fertilizers or even a little herbicide (they could be called "biological"), while others never use man-made fertilizers or toxic chemicals. They rely on the recycling of organic matter and the natural cycles of nutrient flow to supply their crops. They try to build up their soil's humus and improve its structure. They rely heavily on crop rotations, soil organisms and usually composting. They control weeds by non-toxic methods and rely on natural enemies or non-toxic controls for pests and diseases. They shun synthetic fertilizers in the belief that they are harsh and damaging to plants and the soil. Some prefer to plant open-pollinated varieties of crops and avoid hybrids. Organic farmers generally have a high level of commitment and a good understanding of how the soil-plant system works. They are interested in establishing a farming system that is *sustainable*, meaning that it conserves soil and energy resources, maintains a clean environment, produces healthful food, is profitable and promotes a high quality of life not only for the present, but on into the foreseeable future. As we will see later, the definition of an organic grower can also be dictated by the particular organization under which you may wish to become certified.

Organic Farming Methods

There can be different types of farmers lumped under the label "organic." Some may have a deep distrust of anything "scientific," so they may fail to test their soil and eventually have out-of-balance soil, weed and pest problems and low-quality crops. They might be called "organic by neglect." Others are on top of things and keep close track of the status of their soil, their crops and their animals. They don't hesitate to test out new developments and adapt useful methods to their farm system.

To make natural or organic methods work, I think a farmer does need to have more motivation than getting the high market prices. You really need to understand the interrelations of soil, humus, microbes, earthworms, roots, weeds, pests, food quality and animal health. It helps to have a deep regard for the environment and a sense of stewardship of the land. Many biological and organic farmers were motivated to change from conventional methods by concern about the harmful effects of toxic chemicals. Sometimes it took something as awful as a case of cancer in their own family. There are many reasons to go organic, and there are many benefits.

Don't be fooled by the word "natural." Natural, of course, means found in nature, but not all natural products are good. Some of the strong salt fertilizer materials occur as natural salt deposits that are mined for fertilizer. The worst is potassium chloride (muriate of potash or Kalium potash). Some fertilizer salesmen may promote their blended fertilizer as "natural" or even "organic," when it contains half potassium chloride. Another problem is that synthetic urea is technically an "organic" material because it contains carbon (the chemical definition of organic is any carbon-containing compound). Always ask which source materials the fertilizer is made from.

What Does it Take?

Assuming you are considering going organic and becoming certified, what is involved? What can and can't you do? In recent years there has been a proliferation of organic organizations, some local and others national or international. There may also be national organic standards in some countries. Obviously, you need to contact the organization or organizations you are interested in and get their requirements and benefits. There are directories of organic and sustainable organizations which you can utilize.

Becoming certified means that the food you produce can be marketed under a label or logo that signifies a certain standard or quality to the consumer. Most certifying organizations have marketing aids or cooperative marketing arrangements, or perhaps a guaranteed commitment to buy from the grower. Sometimes you can lock in a guaranteed market price ahead of time. But to receive these benefits, you must conscientiously adhere to the organization's requirements or rules and you must be able to furnish proof that your farm is organic. The certification requirements and process typically go like this:

- You must complete a detailed questionnaire concerning your farming methods and the recent history of your practices. Farm records are probably required.
- To be certified organic, a field must have been free of prohibited fertilizers and toxic chemicals for the previous three years. You can certify your farm in portions. You may be required to leave a buffer strip around some fields to catch possible chemical drift from nearby fields. If your land has not been chemical-free for three years, you may be able to qualify as *transitional* and receive above the regular market

prices. Organically grown crops must be harvested, handled and stored separately from any crops grown non-organically.

- You must make a commitment to follow soil-building management methods and show progress in doing so. You must have a written organic management plan.
- You must be able to document the materials you use by an audit trail ("paper trail") of records, receipts or other documentation. Detailed records of dates and methods of application of materials may be required.
- Your farm will be visited and inspected by an organization representative.
- A certification committee will make a decision based on the inspector's recommendation.
- You will have to pay annual dues and licensing fees, possibly determined by your projected volume of sales. You may want to purchase official labels or stickers if you do your own packaging or processing.

Materials generally permitted for use on organically managed soil include most additives of organic matter (animal manures, crop residues, green manures, compost, and various wastes and by-products from animal or plant processing industries), mined natural mineral products (including lime, gypsum, borax and potassium sulfate), and biological products (such as enzymes, hormones and microbial inoculants — see Chapter 16). Some of the above materials may be restricted if they contain toxic heavy metals (such as leather tankage or sewage sludge) or if they could cause soil problems (such as dolomitic lime in high-magnesium soil). Prohibited materials include toxic chemicals (herbicides, insecticides, fungicides and other pest control chemicals), potassium chloride, anhydrous ammonia, ammonium sulfate and monoammonium phosphate (MAP). Some pesticides that are derived from natural sources may be allowed if they cause no environmental harm. There may be restrictions on application of raw animal manures, such as not spreading within 60 days of harvest.

For raising certified organic animals, there are still other rules and restrictions regarding living conditions, feed sources, use of antibiotics and quality of products.

Of course, you don't have to be certified to be organic or to sell your products, but without certification you may not be able to call them "organic" and you may have a harder time selling them for a premium price.

The Transition

Deciding to go organic and successfully accomplishing it are two different things. As with any major change in your life, you should carefully think through and plan what you are going to do. First of all you need to see your goal and know what you want to accomplish. Most importantly, you need to be fully convinced that organic methods will work and be committed to making them work on your farm. Just as with switching to biological methods, you should educate yourself on the principles and natural processes involved in organic agriculture (the contents of this manual would be a good start). You may need to un-learn some previously learned concepts and "re-program" your mind to think in terms of how your operations affect the soil and its life.

Most soils, especially the heavy, high-clay soils, will take some time to get in really good condition, especially if they have been abused. Organic agriculture experts usually say that it will take three to five years to make the transition from conventional to organic. It all depends on the soil type, your climate and what has been done to the soil in the past. If you have already switched from conventional to biological methods, your soil should already be in good condition, and to become certified you would only have to drop prohibited materials and make up for them in approved ways.

It would help many people who may have questions or doubts to visit and talk to successful organic farmers. Most of them would be happy to help someone else along the road to clean farming (of course, respect their need to get their own work done, and do not visit without prior contact). Attending meetings or seminars sponsored by organic organizations is another good way to gain knowledge. There are also many excellent books and magazines which provide sound advice.

Rather than quit conventional methods "cold turkey," most experienced organic farmers recommend a gradual transition, partly so you can develop and fine-tune the new management skills you will need, but also to prevent large losses in case something goes wrong. A good approach is to convert one or a few fields at a time. This will also allow you to use materials "full rate" rather than applying low levels, which may not cause much soil change at all.

At first, do not be surprised if some crop yields do drop below those obtainable with conventional methods. This may be due to problems that develop or mistakes you make while still "learning the ropes." An example is weeds that get out of control. Another possible reason for lower yields is that the yields produced with conventional methods are "empty," inflated yields — crops with a large bulk (bushels or tons) but low quality (nutritional value, test weight). For some crops, quality pays off better than quantity, such as better animal health and productivity from feeding nutritious crops. Organic and biological methods are almost always more profitable than conventional methods in the long run. Using much lower inputs to grow slightly lower yields of higher quality food really pays off. And the good news is that in many cases, yields do not drop much below what you were previously getting. Using the advice of another organic farmer or a consultant will often prevent financial problems during the transition period.

If you have not been using a good crop rotation, or if your soil is hard or heavy, expect some difficulty with weeds at first. Many bad weeds grow best on compacted or crusted soil, and without herbicides you may be swamped by weeds. Review Chapter 19 on non-toxic weed control. Timeliness is very important for successful cultivation and rotary hoeing of weeds, and a spell of bad weather may delay your control efforts. If weeds grow with a vengeance, you may need to weigh the financial consequences of a failed crop and decide to use herbicide and delay your progress toward becoming certified for a year. After your rotation is established and your soil structure begins improving, weed problems should decline.

Another frequent problem transitional farmers run into is providing enough nitrogen for a growing crop when they can't use the conventional nitrogen fertilizer materials. Basically, you have to "grow" your own nitrogen fertilizer. It is a great advantage to have a source of animal manure or compost (or other high-nitrogen organic matter), since this is a major way to recycle and supply nitrogen (see Chapter 18). The other major way to increase soil nitrogen is to grow legumes, which have nitrogen-fixing root nodules. Incorporating a growing legume crop (as a green manure) can add 100 to 200 pounds/acre of nitrogen (be sure the seed is inoculated with the proper strain of bacteria). Some free-living bacteria(not found in root nodules) and blue-green algae can also fix some nitrogen, but not as much as legumes (see Chapter 16, "microbial inoculants").

It is a good idea to keep detailed records of what you do and what happens during the transition period. Of course, record keeping is useful any time, and organic certification requires certain record keeping. But more than just writing down how much of what material you applied, also keep track of important external conditions, like the weather, rainfall and temperature. Better yet, take a little time to check and record such things as the soil temperature and moisture conditions six-inches deep, or the time when earthworm castings appear on the surface. All of these factors can greatly influence later problems (or successes), and if you can check back in your records and figure out what was the cause, you can speed up your learning process. Taking snapshots of significant things you notice can also provide a good record of your progress. Keep track of dates in all of your record-keeping.

Successful organic farming is achievable. Many farmers worry about failures, crop losses and what the neighbors might think if things look less than visually ideal. But our understanding of what it takes and what tools are needed to do the job are here now for you to see. Go visit some successful organic farms. Ask questions, search it out before starting. At this point in time, the financial rewards are strong. Is it a marketing decision? Would you do it if you received no bonus for the products produced? It is not a get-rich-quick scheme. It will take better management and knowledge to be successful.

23

PROFIT PRACTICES

Profit practices are ideas — practices — that farmers are using to either reduce their cost of production, increase yields or increase the quality of their crops, livestock and soil. The last area, quality, is harder to put a value on, but in the long run it is one that really improves profits. It includes healthy plants and reduced chemical use. When *quality* is in the plant, *quantity* (yield) always follows. The soil improvements reduce erosion, improve the crop's ability to tolerate stresses, and maintain and improve the soil's fertility and tilth without the constant physical (tillage) and chemical (nutrients) inputs.

I hope some of these profit practices can be incorporated into your farming system. If you are determined to improve your situation, *don't* use the words, "Oh, I can't." Instead, you must come up with a way that you *can.* Start slowly, one step at a time. Try some of the methods you learn. Working out ways to accomplish "profit practices" is what makes farming fun and profitable.

1. *Test and balance soil* as a guide for what crops to grow and how to apply fertilizers, lime and manure. The lower-testing fields should receive the manure and should not be left in hay very long. Forage crops are the ones that remove a lot of nutrients from the soil. Match your starters, plant food and corrective fertilizers to soil needs. Soil balancing creates an ideal home for soil organisms and a "balanced diet" for crop plants. Adding fertilizers above soil balance requirements wastes money and causes problems. Balanced soils will produce high quality, nutrient-bal-

anced feeds. More than N-P-K must be used. When the major elements, plus calcium, magnesium, sulfur and the trace elements are balanced, profitability improves because crops grow better. Soil testing is a must — then you know where you are.

2. *Fertilizer use and placement.* Split applications and row side dressing or banding of fertilizer (especially nitrogen) gives equal yields and better quality than one-time, pre-plant applications, with less environmental pollution. Often less than one-half the usual rate of nitrogen can be used. A split application of nitrogen can be applied when you cultivate. Seed-applied and in-row plant foods and growth stimulants are very effective and efficient. Row-applied fertilizers allow you to grow a good crop on low-fertility ground with a minimal input.

3. *Seed use.* Soil structure, chemical carry-over, soil calcium level and earthworm activity (plus other soil organisms) all affect seed germination. Twenty pounds of alfalfa seed per acre is 100 seeds per square foot. Twenty-five plants per square foot is a very good stand, so high seeding rates are unnecessary. Less expensive alfalfa seed can do well in good soil. Also, look at lower priced smaller seed corn companies' products, and open pollinated corn.

4. *Run a short rotation.* This is one of the "rules" of biological farming (see Chapter 2). It allows you to drop insecticides, reduce herbicides and incorporate more green manures into a growing cycle. Crop rotation allows you to grow your own fertilizer (legumes add nitrogen for corn), plus it improves crops and reduces chemical input (you get better weed and pest control).

5. *Cover crops.* Anything that you can fall-seed or interseed into an existing crop can make a good cover crop, including rye, vetch, clovers, oats, buckwheat, etc. Not only do these crops protect the soil from erosion, they protect the soil life and grow food for them. They increase and save soil nutrients, and their benefits far exceed the N-P-K value put on them.

6. *Root systems.* Keep in mind your crops' root systems and what will help them. Roots need air and warmth to grow and function. Soil balance with soil life creates looser soils and therefore stimulates root growth. Kelp and humic acid are known root and bacterial stimulants. If you double the root system, you greatly increase the water and plant nutrients taken in by the crop. Chloride and ammonia from commercial fertilizers may burn roots, and heavy nitrogen applications early in the season *decrease* root growth.

7. *Pests and weeds.* Vigorous, healthy plants are naturally immune to pests and diseases. Diseases and insects are "nature's clean-up crew," elim-

inating the sick and weak. Weeds grow best under certain soil conditions. Improving soil structure and balance reduces weed problems.

8. *Herbicide use.* Crop rotation, banding herbicides, adding humic acid to herbicides, using a rotary hoe or cultivator, using nitrogen or sulfuric acid to burn in-row weeds, and flame cultivation are some methods which greatly reduce chemical weed control. Banding can be done with either dry or spray herbicides when planting row crops, or with an early cultivation system. Contact herbicides are good in that they have a lower carry-over and you are selectively putting them on only when needed. Getting your soil well aerated, balancing nutrients and promoting soil life will discourage most weeds (see Chapter 19).

9. *Insecticides.* Crop rotation and healthy plants growing on "healthy," balanced soil make the need for insecticides minimal or obsolete. Continued reliance on insecticides simply creates pesticide-resistant insects and pollutes the environment.

10. *Quality crops and forages.* A good "healthy" soil with balanced fertility and lots of organisms makes for high quality feeds, with good mineral balance, high sugar level and good quality protein. All of these reduce farm costs for protein and mineral supplements, along with animal health care. This increases profitability — the bottom line. For top-quality alfalfa, you need a complete balance of all soil nutrients, including calcium, sulfur and boron. Put excellent quality, ideally stored forages in your silo or mow. Inoculate them to improve storage and pH; make sure there are *no* molds.

11. *Set up a controlled grazing system* to graze pasture land more efficiently and to keep high quality forages growing. Fertilize pastures for added production and quality.

12. *Feed livestock quality minerals* with lots of extras: probiotics, yeast, kelp and B-vitamins. These are things that keep livestock healthy and productive.

13. *Dairy calves.* Many farmers fall short here. Sick calves equal sick cows. They need balanced diets, good nutritional supplements and ideal living conditions.

14. *Dry cow care* for dairy farmers. I always say, give me your dry cows and give me your alfalfa. That's how you make money. Dry cows need special feeds, not left-over junk. *Good* grass is the best dry cow feed.

15. *Livestock manure management.* Storage and handling are part of it. Another part is application. A light coat on many acres is a better use of the nutrients. Heavy doses can do as much damage to the soil and environment as they do good. Stockpiling and/or composting are excellent ways to save livestock manure nutrients. Adding a rock phosphate not

only helps save the nitrogen in the manure by making ammoniated phosphates, but also reduces odors and makes a convenient way to add phosphorus to the soil. With liquid manure, adding rock phosphate (but not the colloidal type of phosphate) minimizes crust, flies and odor. Manure should be tilled into the soil, but keep it near the surface so soil organisms can convert it to plant-usable forms (see Chapter 18).

16. *Corn stalk management.* Right behind the picker or combine, apply some nitrogen or manure and get the stalks mostly worked into the soil. Stalks rot down faster that way. Management of decay is essential for nutrient release and insect control.

17. *Tillage.* Keep these objectives in mind: use tillage to manage organic matter decay and to control the soil's air and water. Avoid compaction-causing operations. A system of tillage will probably include some form of deep tillage occasionally. This will break up a hardpan and aerate soil, although high humus levels and soil life (especially earthworms) will do the job permanently (see Chapter 17).

18. *Erosion.* Compacted and "dead" soils erode easily. Most of your soil's fertility is near the top. Proper management, ground cover, good soil structure and keeping toxins off — all greatly reduce your chance of having compacted soil. Building humus levels, thereby providing better water absorption, will reduce run-off and erosion, as well as increase fertility.

24

HEALTH: HOW TO MEASURE IT

Health

We all want to be in good health. It's no fun to be sick. And then there are the economic consequences: lost wages and production. What is health? How can it be measured and how can it be obtained?

The dictionary defines health as soundness or fitness of physical condition, or well-being. These definitions do not tell how health can be measured or give any clear boundary between health and "non-health." How can you tell if you are completely healthy? What if you mostly feel fine except for a stubbed toe? Does "soundness" or "well-being" mean you have to be at peak potential health?

Other Concepts

Let's consider some other aspects of health. We usually think of total health — our entire body — but actually, total health is a result of the well-being of all the organs and cells in the body. Every cell is like a tiny factory, taking in necessary raw materials (food, oxygen), producing a "product" (perhaps secreting digestive enzymes, or producing nerve impulses) and releasing waste materials (carbon dioxide and other metabolic wastes). Just like a factory, cells run on energy (from food molecules), and they can be running at top production level or somewhere below it. Like a factory, a cell has complex "machinery" to do the work: all of the various cell parts - the nucleus and genes, mitochondria, ribosomes, membranes and so on. The "workers" are the enzymes that assemble raw materials into finished products, and the "boss" or the "brains" of the factory is the DNA stored in the genes.

It's all very complicated. Each cell is a wonder of complexity, and in the whole body, our billions of cells all interact and rely on each other for their needs — somewhat like all the different factories and businesses in a large city. As an entire city can be crippled by a garbage collectors strike, by severe smog or by a hurricane, our body can become sick from low insulin production, from cigarette smoking or from an injury.

Besides the food we eat, the air we breathe and avoidance of accidents, health also depends on our mental condition and on interactions with other people — a low level of stress. There is evidence that we go through regular cycles of ups and downs, called biorhythms.

Not Just People

So far we have been talking about human health, but of course, any living creature can be healthy or not. Those in agriculture are probably more aware than most that animals can become sick, and if they are not healthy, milk or meat production goes down. Just like humans, animals need nutritious food, clean air and water, and a stress-free environment.

But have you ever thought about your crop plants? The same principles apply. To grow well and produce wholesome, nutritious food, plants need water, sunshine, a "balanced diet" of soil nutrients and freedom from storms, drought, toxins and other stresses. It is interesting that just as humans or animals have immune systems and other defenses against diseases, healthy plants also have natural mechanisms that protect them against attacking disease pathogens and insect pests (see Chapter 20).

We could even go one level further and speak of the health of the soil. A "healthy," fertile soil is necessary to grow healthy crops. A healthy soil has adequate air and water, balanced levels of essential plant nutrients, plenty of organic matter and a thriving population of the beneficial organisms that help maintain soil health and nourish plants. These include certain bacteria, fungi and earthworms. Soil is just as complex as a plant or animal or human with regard to its proper functioning. "Sick" soil can cause severe problems just like a sick cow or a sick child. You can use the "Soil Scorecard" (Appendix A) to see how healthy your soil is.

A Complete System

We can tie everything together by noting that all parts of nature, including humans, are interconnected in an ecological cycle, or ecosystem (see Chapter 3). Plants need soil nutrients. Animals and humans eat plants and animals, and produce wastes or organic matter. When wastes are returned to the soil, soil organisms decompose them, and the nutrients nourish plants. Sick or unbalanced soil harms plant health. Sick plants contribute to sick animals, and human health suffers when our food is not of high quality.

How to Measure?

Considering all the complexities of body and cell functioning, it is really almost impossible to accurately measure total health. Do you measure at the body-level or the cell-level? What about a half-hour later? . . . it is probably different then. You might have to measure a hundred or more factors. An accurate, detailed measurement of health probably isn't worth the trouble.

Perhaps a simple way to gauge health is by productivity. If we are working above a certain expected level, we are "doing OK." If our cows give more than 20,000 pounds of milk and our corn field produces 175 bushes per acre, we are pleased. But could they do better? Well, probably yes. If the weather had been warmer or the silage not so moldy, health would have been even better and production higher. *Health is relative*, and most of the time humans, animals and plants — the soil, too — are not operating at peak capability.

We can picture it by a graph, with a "health line" (100 percent health), a "sick line" (when clinical symptoms appear), and a "dead line." Most of the time we, our animals, our crops and our soil operate somewhere between the health line and the sick line — seesawing up and down as we respond to daily and weekly stresses and variations. On "good days," we feel great and can "take on the world," but on poor days we may feel "out of sorts" or "run down." We may have minor, subclinical symptoms of illness. Disease pathogens ("germs"), which are around all the time may start to get a foothold. With proper nourishment and a good night's sleep, we can "shake it off" (our immune system does its job), or we may slip below the sick line and develop clinical symptoms of sickness (fever, sore throat) and spend several days too sick to work. If we don't get rest and proper nutrition, we may get worse or develop pneumonia, TB or some other serious disease. If our vitality continues to sink, we may reach the dead line — and it's all over.

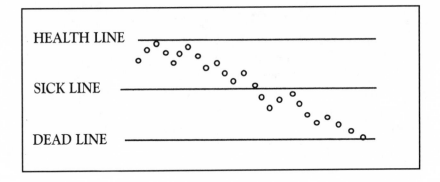

Obtaining Health

Good health depends on many factors. Clean air, pure water, avoiding accidents and toxins are only some. One of the most important is good nutrition. If the "raw materials" contained in food are not of high quality and in adequate amounts, our cellular "factories" will put out an inferior product, or even break down. A proper balance of carbohydrates, fats, proteins, vitamins and minerals is essential for good health. Of course, the details are more complicated: we need complex carbohydrates, unsaturated fats, HDL ("good") cholesterol, essential amino acids, antioxidants, high fiber and so on.

They should be in our food, but are they? What else is in our food that we don't need - like pesticide residues? How was our food grown? With what fertilizers? In what soil? If we can't be sure, we may end up taking a host of vitamin supplements. Most farmers supplement their animals' rations with vitamins and minerals because laboratory tests of their feed show deficiencies. The nutrient content of plants is greatly influenced by the soil in which they are grown. Everything from the amounts and balance of major and minor elements to the amount of air and humus to the number of beneficial bacteria and fungi to the kind of fertilizer materials used can make the difference between nutritious, health-promoting food and "junk."

The way we treat our soil, crops and animals — the way we fertilize and nourish them — is extremely important. Improving soil health contributes to improved plant health, animal health, and, ultimately, to our own health.

25

LIVESTOCK NUTRITION

Although all the information in this book is essential for biological live-stock production, it would be difficult for me not to include a separate chapter discussing nutrition, especially dairy nutrition. After all, I *am* a dairy nutritionist by training, was raised on a dairy farm, and have stud-ied, consulted over and followed cows around for over 25 years after my university training. Trying to grow better feed for dairy cattle is what brought me to study soil mineral balance and soil health. If you have live-stock, you get paid threefold for tending to your soil: once for the improved crop production, again for your improved livestock health (fewer veterinarian visits), and yet again for improved beef or milk pro-duction. When your soils are receiving a balanced diet of nutrients and getting healthier, you can feed your cows following different ration bal-ance "rules." I promote putting cows on a high-roughage diet. Feeding more mineralized, high quality forages allows you to cut down on miner-als, energy and protein supplements. A high-forage diet not only makes for healthy, high-producing cows, but you spend less and earn more.

To fix your soils, change the feeds you grow, and then not take advan-tage of these feeds in the barn is a real mistake. You have to learn to watch your livestock. They will give you clues as to what they want. Besides growing and handling feed for top quality, I like to supplement with a high quality, balanced mineral/vitamin mix. Then supply a free-choice system so you can monitor the cows' desires. I like to put out three mixes: one a kelp/sea salt blend, another a buffer blend, and also a calcium and phosphorus mineral blend. If the cow changes mineral consumption rates, maybe the ration needs to be re-evaluated. The last addition I like in a dairy ration is the "extras" — yeast, kelp, probiotics and chelated trace minerals. The supplements you feed are adding to your livestock's improved digestion and health.

Provide the livestock with comfortable conditions, give them fresh water, high quality feeds and do all the things you can to make them healthy, and they will produce.

In a book titled *Fertility Pastures*, by Newman Turner, written in 1955, the author explains pasture fertility. His definition is that fertility cannot be measured quantitatively. Any measure of soil fertility must be related to the quality of its produce. In a few words, the most simple measure of soil fertility is its ability to transmit, through its produce, fertility to the ultimate consumer. Any breakdown or deficiency in the animal or human is a measure of the infertility of the original soil from which the animal or human derived its food. High production without any supplements certainly represents soil fertility.

In the dairy industry today, high production records are achieved. The dairy nutritionists keep on adding to the soil-produced feeds: bypass protein, chelated trace elements, anionic salts, fats, amino acids and lots of minerals, to mention a few. Turn to the field, supplement there and the healthy foods produced will eliminate many of the expensive supplements. Your cows will be healthy, happy, and produce more!

26

PUTTING IT ALL TOGETHER

If you have waded through all the chapters in this book, your head may be swimming with all the concepts and principles involved in biological agriculture. If these things are new to you, you may be somewhat confused or bewildered. I always say, "confusion is the first step in learning!" The more you study, the more successful farms you visit, and the more you become involved, the simpler it becomes.

The Biological Approach

My approach has been to focus on the "big picture" — to see how the whole natural system works, whereas many experts tend to specialize on some tiny part of the system. They may not appreciate nature's capability to produce bountiful, high quality crops and healthy, productive animals.

Once we have some basic understanding of the whole system, we can much more intelligently decide what to do in our day-to-day farming activities. We can know the best type of tillage to do by knowing what it will do to improve organic matter decay and soil structure. We can decide which kinds of materials we want in our fertilizers based on their lack of harmful effects on crops or the soil.

There are just too many types of farming — too many kinds of crops and animals, too many types of soil — to be able to give the reader detailed, "cook-book" instructions. You have to make your own decisions and work out a system of farming that fits your own farm, based on the biological principles we have covered. An informed decision is much more reliable than one made without having all the important information. That is what I hope I have done in this book — I hope I've given the reader enough background and insight into how biological systems work to be able to make intelligent decisions.

Biological farming does require more thinking than conventional farming, where you may rely on someone else to tell you every little thing to do. Once you begin to understand *why* you are doing what you're doing and you begin to see some real positive changes in your soil, crops and animals — you *know* that you're on the right track. Farming becomes fun.

Steps to take. To review some of the main points, your first step in switching to biological methods is to be somewhat convinced of the "rightness" and potential of the biological approach — at least enough to really try it and make it work. In other words, have a positive attitude. If you haphazardly "just try it" and don't really think it will work — it probably won't.

Then you need to learn all you can about biological principles, to understand what is going on in natural systems. Read, attend seminars and talk to consultants and successful biological farmers.

Decide on your goals — what you want to accomplish. To do that you probably have to analyze your present farming system and pin-point the problems that need fixing. Look at everything: soil structure, nutrients, tillage methods, crop rotation, fertilizers and liming materials, planting and harvesting, weed control, manure management and so on. Do whatever soil and crop testing you need in order to find out where you are now (it is money well spent; stabbing in the dark is futile). Get the advice of an informed consultant if you need help in tracking down problems.

Formulate a specific plan to reach your goals, covering say, five years. Change your crop rotation if necessary. Plant more cover crops and green manure crops. Begin using better fertilizers. Do whatever tillage is necessary to get improvements in soil structure started, then encourage beneficial soil organisms so they can take over the job. Try to cut down on toxic chemical use, but gradually so as not to lose a crop. To take 10 acres and try this thing called biological farming may be just wasting your time. You need to have a complete program for change. Using high quality balanced fertilizers and switching to some of the practices mentioned in this book is not risk. You can just trade dollars, since in most cases farmers can't afford to spend more. If extra spending for soil correction is required, work within your budget. The returns will come.

Work out a budget to substitute biologically friendly fertilizers and materials for your old conventional inputs (spend whatever money you have available on different things). Again, the advice of a consultant can really help in the transition period.

Plunge into it with high expectations, and make it work. Do the best you can, and don't let minor problems side-track you. Instead, find a way to solve them. Expect some difficulties until you get comfortable with your new methods. Making changes in any aspect of life can be difficult — and worrisome at times. Having another biological farmer or a consultant to talk to can help a lot. Remember that natural systems are designed to work out fine. Just do the basic things we have covered in this book and the soil organisms and plants will take it from there.

Once you begin to see good things happening — like the return of earthworms — you may want to "convert" all your neighbors and relatives, but don't be surprised if you get a cool reception. Not everyone else can see the desirability of following natural methods, or they may be skeptical, or afraid to change, or afraid of what their neighbors would think. Sometimes the best way to convince others is by example.

To save farmers and change agriculture to a more profitable, environmentally friendly system takes many successful biological farms. More farmers and educators are asking questions and watching. A lot of the soil scientists, agronomists and nutritionists do not understand what biological farming is. Some may have a part or piece, and some may speak against it. Perhaps you feel you can't do this type of farming. Maybe the "can't" should be a "won't." "I won't grow green manure crops. I won't rotate crops. I won't till." Don't lock yourself in a box — do the job. Maybe because of these won'ts you won't get the soil and crop to perform and your farming operation won't be sustained. If things are slowly getting worse and you do nothing about it, it is only a matter of time before change will be forced on you, or your children if they are to farm.

If you would like guidance, look for a consultant who is knowledgeable and willing to work *with* you to solve problems and give suggestions. A good consultant does not just sell products; he/she educates and informs. When a consultant comes to your farm, he/she should work with you on farm goals, pointing out things about your soils, crops and livestock — both the positive things and areas that may need improvement. A consultant should work *with* you to facilitate changes beginning from where you are right now. Making changes too quickly is not always ideal. If you have questions, a good consultant will seek out answers, and report to you on farm demonstrations or research. If you are to spend time and money with a consultant, these are not unrealistic expectations.

Summing it up. As we have done in Chapter 2, we can briefly sum up the principles of biological agriculture in the six "rules of biological farming."

Rule 1: Test and balance your soil.

Test to see where you are and where you need to go. Then apply the materials necessary to move towards a better balance of *all* plant nutrients.

Rule 2: Use fertilizers which are life-promoting and non-harmful.

Depending on what the soil needs, use materials that will give good plant growth without harming the plant, soil organisms or soil structure.

Rule 3: Use pesticides and herbicides in minimum amounts, only when absolutely necessary.

Work toward building up soil "health" so that noxious weeds will decrease and crop plants will be healthy and will naturally resist pests and diseases.

Rule 4: Use a short rotation.

A short rotation lessens weed, pest and disease problems, and can provide some crop nutrients and improve soil structure.

Rule 5: Use tillage to control decay of organic materials and to control soil air and water.

Good soil structure and decay of organic matter are integral parts of a biological system. Use tillage to help nature along.

Rule 6: Feed soil life.

A healthy population of beneficial soil organisms is your best ally and a necessary part of growing a good crop and maintaining good soil. Keep them happy.

Appendix A

SOIL SCORECARD

Every farmer wants to have the best soil he can. Fertile soil with good structure is more than half the battle. Crops on good soil will grow better and be healthier, with fewer disease and pest problems than stressed crops grown on poor soil. They will withstand drought better and produce better quality forage or food. Livestock will be healthier and more productive when they eat more nutritious feed. All in all, good soil helps improve your profitability. You need to add less fertilizer and use less pesticide, so the environment is helped, too.

How can you tell if your soil is in top-notch condition, or just mediocre? Soil scientists could do complex and expensive lab tests, but that is impractical. Just by making several simple observations of your soil, crops and animals, and using some of the figures on your soil tests, you can get a good idea of your soil's condition.

There are many types of soil. Soil is very complex, and growing top quality crops depends on many soil factors. When you add to those the changeable quality of the weather, it's not hard to see why growing crops is so uncertain. You can't control the weather, although good soil conditions will partly overcome adverse weather effects, such as drought, frost and waterlogging.

We can group the various soil factors or characteristics into three large groups: physical, chemical and biological.

Physical refers to soil structure and texture, its density and particle sizes. These relate to the soil's tilth, or looseness and ease of tillage. Good structure gives good aeration, water uptake and drainage, and erosion resistance.

Chemical refers to the soil's content and balance of nutrient elements, such as calcium, phosphorus, potassium, trace elements and so on. We usually call this "fertility," but really that term includes the physical and biological factors also. The soil's ability to hold nutrients (its cation exchange capacity, or CEC) and its pH (acidity or alkalinity) are other chemical factors of the soil.

Biological means the life in the soil — the multitude of living organisms that live in the soil. Most of them are microscopic, so we seldom think about them. There are good ones and bad ones. The bad organisms include disease-causing microbes, rootworms and some nematodes. The good ones include the bacteria, fungi and earthworms that help decompose crop residues and manure so the nutrients they contain can be used by crops. But the "good guys" do much more. They improve soil structure, make unavailable nutrients more available to plants, and they even help protect plants from diseases and pests. Either directly or indirectly, they help plants grow better and produce better quality food.

Using the Soil Scorecard

The Soil Scorecard is divided into five sections. Each section has a number of questions, each dealing with a soil factor. In each question are two to five choices, followed by a point value, from zero to three points. Answer as many of the questions as you can about your soil or farming practices. You may want to go out into the field to check some. If you are unable to answer the question, or if the question does not apply to your farm, skip that question. Write your point score for each question in the blank at the right. If your soil happens to fall between the whole numbers, you can either pick the closest score or use a decimal score, such as 2.5. Follow the instructions at the end of each section for finding the average scores, and at the end of the scorecard for finding your overall soil rating.

You can use the Soil Scorecard in a general way for your entire farm, or use it to rate individual fields or soil types. Your overall soil rating will give you a general idea of how your soil compares to an "ideal" soil. If your soil has some serious problems, it would help to look at the individual scores to find the low numbers and to possibly trace the cause. By knowing the cause of the problem, you can begin to fix it. Some soil problems may be easy to correct, while others may take years just to begin to fix. Since soil condition involves so many factors, it can be difficult to track down the real cause of a problem, and correcting it may involve changing some of your farming methods.

Section 1 — Physical

(Questions refer to the plow layer, the upper six or seven inches.)

1. Soil Density:

   Soil dense (or hard or compact) most of the time — 0 points.

   Soil dense only in spots — 2 points.

   Soil loose, easily tilled — 3 points.

   Your score _____

2. Soil Hardpan:

   Thick hardpan beneath surface (about six-inches thick) — 0 points.

   Only a thin hardpan (a couple inches thick) — 1 point.

   No hardpan — 3 points.

   Your score _____

3. Soil Condition in Dry Weather:

   Soil cracks or forms a crust in dry weather — 0 points.

   Soil cracks or crusts only in spots — 1 point.

   No cracks or crusting in dry weather — 3 points.

   Your score _____

4. Soil Texture:

   Texture is cloddy or sticky — 0 points.

   Texture is with very fine particles, but soil is dense — 1 point.

   Texture is sandy or coarse — 2 points.

   Texture is loamy and crumbly — 3 points.

   Your score _____

5. Soil Water Absorption:

   Water usually ponds in low spots after a heavy rain — 0 points.

   Water rarely ponds in low spots — 2 points.

   Water never ponds in low spots — 3 points.

   Your score _____

6. Soil Erosion:

   Severe runoff and erosion in moderate to heavy rain — 0 points.

   Some runoff and erosion in moderate rain — 1 point.

   Some runoff and erosion in heavy rain — 2 points.

   Little or no runoff or erosion — 3 points.

   Your score _____

7. Soil Durability:

   Soil very drought-prone — 0 points.

   Soil occasionally drought-prone — 1 point.

   Soil drought-prone only in spots — 2 points.

   Soil not drought-prone — 3 points.

   Your score _____

8. Soil Color:
   Soil color light (white, yellow, light gray) — 0 points.
   Soil color brown or dark gray — 2 points.
   Soil color dark brown or black — 3 points.

   Your score _____

   Now add your eight scores for the Physical section and divide by eight to get your average physical score:

   total physical score = _____ ÷ 8 = _____.

   If you couldn't rate your soil on one or more questions, divide by the number of questions you answered.

Section 2 — Chemical

(Questions refer to the plow layer, the upper six or seven inches, and to soils of temperate climates. You should be able to get the required information from a standard soil test.)

1. Calcium Level:
   Calcium is less than 70 percent of CEC (percent base saturation) — 0 points.
   Calcium is 70 to 74 percent of CEC — 1 point.
   Calcium is over 85 percent of CEC — 2 points.
   Calcium is 75 to 84 percent of CEC — 3 points.

   Your score _____

2. Magnesium Level:
   Magnesium is over 30 percent of CEC (percent base saturation) — 0 points.
   Magnesium is 25 to 29 percent of CEC — 1 point.
   Magnesium is 19 to 24 percent of CEC — 2 points.
   Magnesium is below 12 percent of CEC — 1 point.
   Magnesium is 12 to 18 percent of CEC — 3 points.

   Your score _____

3. Potassium Level:
   Potassium is less than 2 percent of CEC (percent base saturation) — 0 points.
   Potassium is 2 to 3 percent of CEC — 1 points.
   Potassium is over 5 percent of CEC — 2 points.
   Potassium is 3 to 5 percent of CEC — 3 points.

   Your score _____

4. Phosphorus Levels:
   Phosphorus (P2) is less than 30 pounds/acre (15 ppm) — 0 points.
   Phosphorus (P2) is 30 to 39 pounds/acre (15 to 19 ppm) — 1 point.
   Phosphorus (P2) is 40 to 90 pounds/acre (20 to 45 ppm) — 2 points.

Phosphorus (P2) is 91 to 120 pounds/acre (46 to 60 ppm) — 3 points.

Your score _____

5. Sulfur Level:
   Sulfur is less than 30 pounds/acre (15 ppm) — 0 points.
   Sulfur is 30 to 39 pounds/acre (15 to 19 ppm) — 1 point.
   Sulfur is 40 to 49 pounds/acre (20 to 24 ppm) — 2 points.
   Sulfur is 50 to 100 pounds/acre (25 to 50 ppm) — 3 points.

Your score _____

6. Nitrogen-to-Sulfur Ratio:
   Nitrogen-to-sulfur (N:S) ratio over 20:1 — 0 points.
   N:S ratio from 16:1 to 19:1 — 1 point.
   N:S ratio from 11:1 to 15:1 — 2 points.
   N:S ratio below 10:1 — 2 points.
   N:S ratio about 10:1 — 3 points.

Your score _____

7. Iron-to-Manganese Ratio:
   Iron-to-manganese (Fe:Mn) ratio far above or below 1:1 and one or
   both of these elements below 30 pounds/acre (15 ppm) or above 50
   pounds/acre (25 ppm) — 1 point.
   Fe:Mn ratio about 1:1 and both elements about 40 pounds/acre (20
   ppm) — 3 points.

Your score _____

8. Trace Elements:
   Most trace elements (zinc, manganese, iron, copper, boron) test
   "Low" — 0 points.
   One or two trace elements test "Low" — 1 point.
   Most trace elements test "Medium" — 2 points.
   A few trace elements test "Very High" and the rest test "Low" or
   "Medium" — 1 point.
   All trace elements test "Medium" or "High" — 3 points.

Your score _____

9. Cation Exchange Capacity (CEC):
   CEC (cation exchange capacity) below 9 — 1 point.
   CEC between 10-15 — 2 points.
   CEC above 15 — 3 points.

Your score _____

10. pH:
    pH below 5.5 or over 7.6 — 0 points.
    pH between 5.5 to 5.9 or between 7.3 to 7.5 — 1 point.
    pH between 6.0 to 6.4 or between 7.1 to 7.2 — 2 points.
    pH between 6.5 to 7.0 — 3 points.

Your score _____

11. Organic Matter:

  Organic matter tests "Low" or below 1 percent — 1 point.

  Organic matter tests "Medium" or 1 to 2 percent — 2 points.

  Organic matter tests "Very High" or over 5 percent — 2 points.

  Organic matter tests "High" or 3 to 5 percent — 3 points.

  Your score _____

  Now add your 11 scores for the Chemical section and divide by 11 to get your average chemical score:

  total chemical score = _____ ÷ 11 = _____.

  If you couldn't rate your soil on one or more questions, divide by the number of questions you answered.

Section 3 — Biological

(Questions refer to the plow layer, the upper six or seven inches, and during the warm season of the year.)

1. Earthworms:

  With no or almost no earthworms (0 to 5 worms/square foot) — 0 points.

  With a small number of earthworms (6 to 10 worms/square foot) — 1 point.

  With a moderate number of earthworms (11 to 24 worms/square foot) — 2 points.

  With many earthworms (25 or more worms/square foot) — 3 points.

  Your score _____

2. Residues and Manure:

  Previous year's buried residues or manure do not rot — 0 points.

  Slow rotting of previous year's buried residues or manure — 2 points.

  Rapid rotting of previous year's buried residues or manure — 3 points.

  Your score _____

3. Soil Smell:

  Soil with putrid, sour or unpleasant smell — 0 points.

  Soil with no odor, or a "mineral" smell — 1 point.

  Soil with rich, "earthy" smell — 3 points.

  Your score _____

  Now add your 3 scores for the Biological section and divide by 3 to get your average biological score:

  total biological score = _____ ÷ 3 = _____.

If you couldn't rate your soil on one or more questions, divide by the number of questions you answered.

Section 4 — Other Signs

(Questions refer to typical years, when conditions are not related to severe weather, drought, flooding, etc.)

1. Crop Growth:

   Crops grow very poorly, never mature, or die — 0 points.

   Crops grow slowly, mature late — 1 point.

   Crops show nutrient deficiency signs (such as leaves turning yellow, white, reddish, bluish or gray) (not related to pests or diseases) — 1 point.

   Crops seem to lag in growth, but still mature all right — 2 points.

   Crops grow well, mature on time — 3 points.

   Your score _____

2. Crop Pest & Disease:

   Crops with pest or disease problems widespread or frequent — 0 points.

   Crops with pest or disease problems spotty or occasional (about 2 years out of 5) — 1 point.

   Crops with pest or disease problems rarely — 2 points.

   Crops normally with no serious pest or disease problems, or only in stress conditions (drought, extreme weather, etc.) — 3 points.

   Your score _____

3. Crop Yield:

   Crop yields very low — 0 points.

   Crop yields low — 1 point.

   Crop yields moderate — 2 points.

   Crop yields high — 3 points.

   Your score _____

4. Forage or Feed Nutrient Tests:

   Forage or feed nutrient tests low in nutrient value or unbalanced — 1 point.

   Feed tests moderate — 2 points.

   Feed tests high and balanced — 3 points.

   Your score _____

5. Animal Health:

   Continuous animal health problems (unrelated to poor housing, poor ventilation, poor water or bad weather) — 0 points.

   Frequent animal health problems — 1 point.

   Occasional health problems — 2 points.

   No or rare animal health problems — 3 points.

   Your score _____

6. Weeds:
Fields with severe noxious weed problems — 0 points.
Fields with constant but not severe weed problems — 1 point.
Fields with spotty or occasional weed problems — 2 points.
Fields with no or rare weed problems — 3 points.

Your score _____

Now add your six scores for the Other Signs section and divide by six to get your average other signs score:

total other signs score = _____ ÷ 6 = _____.

If you couldn't answer one or more questions, divide by the number of questions you answered.

Section 5 — Farming Practices

1. Tillage:
Often drive on or till soil when it is wet — 0 points.
Occasionally drive on or till soil when it is wet — 1 point.
Rarely drive on or till soil when it is wet — 2 points.
Never drive on or till soil when it is wet — 3 points.

Your score _____

2. Herbicides/Pesticides:
Use herbicides/pesticides every year at full rate for definite problems — 0 points.
Use herbicides/pesticides every year at full rate for insurance — 0 points.
Use herbicides/pesticides only when needed, at full rate — 1 point.
Use herbicides/pesticides only when needed, at reduced rate or banded or with surfactant — 2 points.
Never use herbicides/pesticides — 3 points.

Your score _____

3. Nitrogen Source:
Use anhydrous ammonia as main nitrogen source — 0 points.
Use urea or DAP (diammonium phosphate) as main nitrogen source — 1 point.
Use liquid 28 percent, ammonium sulfate or ammonium nitrate as main nitrogen source — 2 points.
Use manure and/or legume plow-down as main nitrogen source — 3 points.

Your score _____

4. Nitrogen Source:
Use "standard" commercial fertilizers containing urea, DAP and potassium chloride (0-0-60 or 0-0-62) — 0 points.

Use special "low-salt" blends or materials, without manure, green manure or compost — 2 points.
Use special "low salt" blends or materials, plus manure, green manure or compost — 3 points.

Your score _____

5. Soil Tests:
Never test soil, only apply fertilizer/lime by "guess" — 0 points.
Test soil about once every 5 to 10 years; only test P, K and pH — 1 point.
Test soil regularly, at least every three years; do complete test (all elements, including trace elements and CEC base saturations) — 3 points.

Your score _____

Now add your five scores for the Farming Practices section and divide by five to get your average farming practices score:

total farming practices score = _____ ÷ 5 = _____.

If you couldn't answer one or more questions, divide by the number of questions you answered.

Now add your five average scores:

Physical = _____

Chemical = _____

Biological = _____

Other Signs = _____

Farming Practices = _____

Total = _____

Now divide the total by five to get your Overall Soil Rating:

total = _____ ÷ 5 = _____.

Here is how your soil rates:
Overall Soil Rating = 0 to 0.9 = severe problems, needs help.
  1 to 1.9 = needs much correction.
  2 to 2.7 = good, needs improvement.
  2.8 to 3 = excellent.

If your overall soil rating was below 2.0, you can expect to spend three to five years to correct the major problems, provided you make the necessary changes in farming practices and soil additives (lime, fertilizers, organic matter).

If your overall soil rating was 2.0 to 3.0, your soil is in good condition, but will need minor correction or yearly maintenance. It should respond quickly.

Whether you are just beginning or are already changing to biological farming methods, you should see positive changes soon. Abused soils may take two or three years, but you should see earthworms return (in most non-sandy soils) and animal health improve in a shorter time.

Appendix B

RESEARCH:
WHAT IT IS & ISN'T

Most farmers rely heavily on scientific research. You may have read about a research project in a popular farming magazine, where the scientist was quoted giving conclusions based on his research, often with accurate numbers to back him up. Maybe he said that corn yields are "significantly higher" using no-till methods than using moldboard plowing. Does that mean you should stop plowing and never till again? And what does "significantly higher" mean?

What is Research?

You need to understand a little about what scientific research is, the philosophy behind it and how it is done. Scientists use what is called the *scientific method,* a term that refers to the way they search for "the truth." Scientists usually are really trying to find out as much truth as they can about whatever they are studying. Because the universe is so large, most scientists become narrow specialists and restrict their research to only limited subjects, which often seem amusing to the layman (like nitrogen-fixing bacteria of lupines, or the rumen digestion of goats). Sometimes what they study is partly determined by their source of funding (maybe a national goat breeders association was giving a research grant).

In using the scientific method, scientists usually try to answer only a certain specific question, such as the effect of using anhydrous ammonia on the survival of nitrogen-fixing bacteria of lupines (lupines are a type of legume). They may not want to take on too large a project, and it is easier and faster to do such specific research (often small projects are chosen as being suitable for a graduate student's research).

Scientists use certain designs or plans of action for their research that are intended to unquestionably answer the specific question(s) they are studying. In trying to find the "truth," they are seeing what happens when certain conditions exist, and they do experiments using differing conditions, such as seeing what different rates of anhydrous ammonia do to nitrogen-fixing bacteria of lupines; say at 50 pounds/acre, 100 pounds/acre and 200 pounds/acre. But they also want to compare those tests with an untreated situation (no anhydrous ammonia), so they call this the "control" or "check." That way they can tell if any differences are due to the ammonia alone or to other factors also.

To try to eliminate factors that they are not studying, scientists try to keep as many other conditions as possible the same. For example, they may use the same type of soil for all their tests, with identical amounts of other nutrients (phosphorus, potassium, etc.) and identical amounts of water, light, etc. They may do the experiments in a greenhouse with the plants in pots so as to be better able to keep conditions uniform. They would use the same species and variety of lupin plants and the same strain of nitrogen-fixing bacteria in all tests. They would try only to vary the rate of anhydrous ammonia applied. If they were really careful, they would even run the ammonia-injecting apparatus through the control tests also, but with no ammonia being released, just to be sure the soil disturbance was the same for all tests.

In order to be able to draw some definitive conclusions, scientists also do an experiment more than once, usually at least three or four, or as many as ten or more times. This is done to allow them to be more certain of their results. If you flip a coin once and get "heads," does that mean you will get "heads" every time? No. If you flip it twice, you could get another "heads" or you may get a "tails." The more times you flip the coin, the closer you will come to the "truth," which is that you have a 50 percent chance of getting either a "heads" or "tails." What results the scientists might get from doing an experiment only once could be very close to the "true" result, or they could be far removed because of chance variations and possible "experimental error" in measuring. To try and reduce these random sources of error, scientists do their experiments several times (called replications) and they use statistical analysis (a type of mathematical treatment) to tell them how reliable their results are. You would do the same type of thing when you calculate the *average* of several numbers; an average (or "mean" as statisticians call it) is a more reliable number than any of several individual measurements.

By using statistical analysis, scientists can say their results (perhaps an average lupin yield of 21.3 bushels/acre for the control versus 16.8 bushels/acre for 100 pounds/acre of anhydrous ammonia) are "statistically significant at the 95 percent level." This means that the chances are 95 to 5 that there is really a difference between those two average yields due to the treatment (100 pounds/acre of anhydrous ammonia versus no ammonia). A 95 percent level of certainty is the usual minimal standard for research to be considered statistically significant.

We have explained all of the above to show that scientists do their experiments very carefully, so they can be very certain of what is causing the result that they observe. Scientists are usually careful not to overstate their findings. They may say that their results with the anhydrous ammonia and lupines only apply to that strain of bacteria, that variety of lupin, to that soil type and to those greenhouse conditions of temperature, moisture, etc. This search for exactness and certainty is to be commended and it is part of the scientific method. Other aspects of the scientific method include being able to repeat experiments and get the same basic results each time. The search for "truth" by the scientific method is intended to be an unbiased, logical use of observations or measurements, searching for possible explanations (hypotheses), re-testing and formulating probable explanations (theories). Scientists should be willing to abandon or change their theories if additional research shows that they are incorrect.

What Research Isn't

In spite of all the idealistic and impressive aspects of scientific research, it may not always be as reliable as it seems. There are some problems and some things the average farmer should be aware of when he is evaluating research or applying research findings to his farming operation.

1. Most research is very specific. Can you apply the results of a greenhouse experiment on a certain variety of lupines grown in a certain type of soil to field conditions with a different crop variety or species in a different soil type? Maybe or maybe not. Scientists often do follow-up experiments under different conditions to be better able to draw more general conclusions. Research done in Georgia may not apply to your field in Minnesota. You would need to find out the details of a scientist's research, which can usually be done by either finding the original publication in a scientific journal, or by contacting him personally.

2. Realize there is some uncertainty involved in scientific conclusions. A 95 percent probability is a pretty good bet, but the results may be quite different under different conditions (for example, the weather has a way of throwing you a curve just about every year, it seems). A scientist's conclusions only apply to those exact conditions under which his experiment was done. There is so much variability in nature — in soils, plants and animals — that many, or maybe most, of the scientific studies done have little direct application to the "real world." Additional research may find some general principles that can be applied to practical situations. That is where technology (applied science) comes in.

3. Although it shouldn't happen, occasionally scientists are biased. They may design an experiment so a certain product or factor will fail or not be "statistically significant" from the control. That can happen when only a few replications are run or when there is a lot of variability in the measurements. You can "prove anything with statistics" by the way you design an experiment and analyze the data.

4. Sometimes scientists are careless, or they "ask the wrong question." They may not control all the variables. Some greenhouse tests are done with normal field soil (with living organisms present), while in other experiments the soil is sterilized to kill all life. Using sterilized soil may be more exacting for certain experiments, but it has no relevance to actual real-world field conditions where organisms are present (soil organisms can have large effects on plant growth). Some experiments are even done with plants growing on a nutrient solution with the roots in glass beads rather than soil, or maybe just in a nutrient solution. The researcher is trying to eliminate the variables caused by soil, but is he really asking the right question?

A good example is a project at the University of Wisconsin where they are trying to determine if corn starter fertilizers pay. The fertilizer trials are being conducted at 75 sites around the state. They are measuring corn height and grain yield. The fields selected were excessively high in P and K, and the starter fertilizer was a 10-20-20 N-P-K commercial type. At the 75 sites, only a four-bushel average yield increase was obtained. This only profited 40 percent of the users. Does this study conclude starter fertilizers don't pay? With soils excessively high in P and K and adding a fertilizer which only contains N-P-K, we shouldn't expect to get a response. Adding something you already have too much of probably won't benefit you. If these soils are excessive due to high manure use, I would suggest a better manure management program. If they are excessive due to high commercial N-P-K use, back down and find other soil limitations. Just because P and K are high does not mean there are not other limiting nutrients. What about the quality of the crop? What about nitrogen; how much of that was added? Research is based on the questions we ask. In this case, the answer seems obvious without all the expense of research. If your soil P and K levels are excessive, adding more will probably not provide any benefit — case closed.

There are many other questions about the research area called starter fertilizers that could be asked: types of fertilizers and effects on plant mineral uptake, root development, soil life, soil structure and weed growth, to mention a few. Some of these things are hard to measure. Maybe that is why the research is not being done.

Almost all agricultural research is done by measuring only the yield of the crop (perhaps dry weight, or wet weight, or length of stalks, or bushels per acre). Such quantity measurements are relatively easy to make, but is the scientist asking the right question? Is bulk more important than quality in a crop? Maybe it is sometimes, but most crops are grown to feed animals or humans. Their nutritional value would seem to be a relevant thing to study. Yet very few scientists consider the big picture. They are too specialized, limited in their narrow specialties. Often, agronomists or soil scientists never talk to plant pathologists to see why diseases attack crops, or they have no idea what their fertilization recommendations do to the animals that eat the crops (this is an actual situation; high-potassium fertilization of forages leads to low calcium and low magnesium levels in livestock diets, causing serious health problems).

So much of agricultural research is simply impractical or out of touch with the real needs of farmers. It also tends to be biased toward the conventional, chemical-intensive methods that have been tried since World War II. Low-quality food, pesticide residues and polluted ground water are a few of the results of short-sighted scientific research. The fault isn't really with research or the scientific method as such, but with the biases and direction in which research has taken agriculture. Much research is funded by industry, and of course the overriding motive of industry is to make a profit — a *big* profit if possible. Sadly, the welfare of people and the environment usually take a back seat.

Agricultural research has often failed to recognize the importance of the natural laws and systems that regulate the functions of the soil, plant growth and animal performance. More often, the philosophy of science and industry has been to overpower or replace nature. But Mother Nature bats last. Natural systems and the delicate ecological balance of nature can be upset, and upset natural systems can wreak havoc.

Rather than relying on a specific research report (not that all research is false or bad), carefully evaluate the latest hot-off-the-press announcements and products to see if they have any real value or whether they might somehow result in some long-term environmental problem. Is it cheaper and safer to let natural organisms do the job instead? Is it best for your dairy cows to give them daily injections of a synthetic growth hormone so that they will give a little more milk, when just by feeding them high quality feed they will do the same thing, and remain healthy? Keep the big picture and the long haul in mind. Will your farm be worth leaving to your grandchildren?

Do Your Own Research

Many of the practices of biological and organic agriculture had their origin with farmers, *not* the academic world. Practical, "can-do" farmers have come up with ideas, tried them out, modified and perfected them. Such practices as organic matter recycling and controlled grazing are largely farmer-developed. Being closest to the land and having to deal with day-to-day problems, it is only natural that this would be the case. With a basic understanding of the principles of natural systems, a vision of goals to accomplish, and a lot of common sense and perspiration, farmers can do as much or more than academia to help sustainable types of agriculture grow and become the "conventional agriculture" of the future.

If you want to really get a good idea of how a new variety, a different fertilizer or a growth-stimulating product will perform on your farm, take the effort to do some more careful testing. You probably won't want to get so involved that you are doing the type of research that scientists do, yet with only a little extra trouble you can do some meaningful experiments — you can do your own research. On-farm research is becoming more common with some universities (you may be able to get some professional advice and help with expenses), and there are some farm organizations that help farmers do on-farm research. Some states or organizations give grants for farmers to demonstrate sustainable methods. Being able to show other farmers that biological farming actually works on your farm goes a long way toward furthering the cause of environmentally sound agriculture.

When doing your research, run at least three (or better yet, four) duplicate tests (or replications). This way you can get an average and know with more certainty what effect the treatment you are researching had. Try to keep other variables as constant as possible. Use as uniform a field as possible, with little variation in slope, soil types or fertility. You will be setting up several test plots. Their size should be large enough to allow easy planting, cultivating and harvesting with your equipment, but fairly small so as to reduce the effect of field variability. You might use narrow strips across a long, narrow field (see illustration). Often professional researchers use small, garden-sized test plots, but those usually require a lot of hand labor or special small-scale machinery.

For many tests, you should use the center of each test strip (say four or eight rows) as the actual crop to be sampled (harvested or tested) and leave a narrow border or edge of the same crop (say four rows) around the edges, but not harvested for the sample. If there is not likely to be an "edge effect" due to drift, this can be eliminated.

To keep it simple, at first only compare one or two treatments; such as two different nitrogen fertilizers (or the same fertilizer at two different rates). You also need to compare the treatment(s) with a control, or no treatment. So, if you are comparing two fertilizers, you would have those two tests plus a control, or a total of three tests or comparisons (you could of course do any number). You should also run three or four replicates of each test, so you will wind up with as many as a dozen separate plots or strips. It is important that the actual place in the field where each strip is located be determined at random, to eliminate any possible bias due to non-random choosing. It may be best to group the three (in this example) test strips for each set of replicates together into a test block (each block would have a control plus the two treatments), but then to locate the position within a block at random (see illustration). You can use coin-flipping, dice-throwing or any such random method to designate the position of the strips (for example, you could say that the control will equal "1" or "2" on the dice, fertilizer "A" equals "3" or "4" on the dice, and so on).

After mapping out the test strip locations, take care to do the product application, planting, and any other operations as uniformly as possible on all strips. During the growing season, check the strips often to see if there are visible differences. Write down any observations, take measurements, perhaps do tests (maybe plant tissue tests, or refractometer tests) and take pictures so you will have better proof of results.

If the experiment requires harvest results, be sure to take the time to get accurate yield checks (or other tests). Quality measurements may also be useful (test weight, moisture content, protein, etc.).

You will want to do at least some data analysis. Is there a lot of variation within the same treatment? If, for example, yields of the four replicates for the same treatment range from 75 to 150 bushels/acre, the results are much less reliable than if they were all within five bushels of one another. Calculate averages of the replicates for each treatment, and compare them. If there is little variation within treatments, if there is little or no overlap in the ranges between different treatments, and if there is at least a 10 percent difference between two averages, it is likely that the two averages are "statistically different." You can do an actual statistical analysis with the

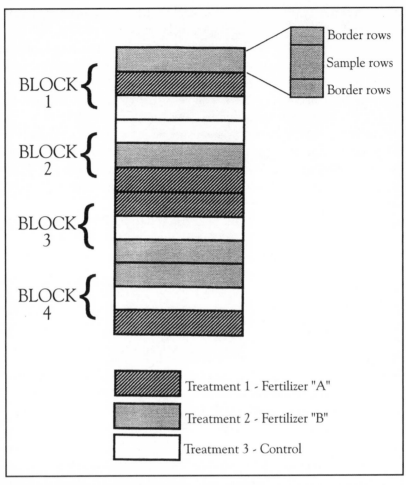

Border rows
Sample rows
Border rows

BLOCK 1

BLOCK 2

BLOCK 3

BLOCK 4

Treatment 1 - Fertilizer "A"

Treatment 2 - Fertilizer "B"

Treatment 3 - Control

aid of a textbook, a math teacher or a computer program. For example, let's say your test results came out as follows (in bushels/acre):

	Treatment 1	Treatment 2	Treatment 3 (control)
	141	156	129
	135	172	139
	146	165	119
	144	181	137
average =	141.5	168.5	131.0
range =	146-135	181-156	139-119
=	11 bushels	25 bushels	20 bushels

At first glance, it would seem like all three averages are quite different from each other, but note that there is a little overlap in the range of figures in treatments one and three (the lowest yield of treatment one overlaps the highest yield of treatment three). If you subtract 10 percent from 141.5 or add 10 percent to 131.0, you have to conclude that the two averages are probably not significantly different. Certainly the average for treatment two is more than 10 percent different from the other two averages and their ranges do not overlap, but note there is more variability in treatment two (a 25 bushel range), so this average is not as certain as the others (treatment one has much less variation than the others, so it is the most certain). In this hypothetical example, we can say that treatment one probably had no yield effect, while treatment two almost certainly did. A "real" statistical analysis would give you exact comparisons and let you know for sure which treatments were significantly different.

Repeating the tests the next year would eliminate possible variations due to weather and give you more certainty in the results.

Doing your own testing or research can take some extra time and trouble, but it is the best way to compare products or practices to see how they work on *your* farm.

Otter Creek Farms — Seaweed, Fish and Humic Acid Demonstration Study:

It makes sense (and is a lot cheaper) to make an argument for a theory that is shown to be effective after many repeated demonstrations than it is to actually conduct the research. If one farm sees results from a certain treatment, maybe it is just an accident. Ten farmers doing a similar thing and getting similar results certainly gives more confidence to the idea. Now, go to a hundred farms. If a majority of these farms generate a positive response from comparable treatments, you now have my attention. If one feels it is necessary, research could now be done to give further confidence to the "doubting Thomas's" who need scientific research.

A study to demonstrate the effectiveness of foliar feeding, a seaweed extract, fish and humic acid blend was outlined for our farm. We conducted this study based upon comments received form others that have benefitted by using these products. Replicated treatments were set up for a soybean foliar to be applied at pod set on the beans. We conducted our experiment in a field where a lot of mineralization, green manure crops, composts and other soil-building practices were already done. Along with this, a good, balanced fertilizer was used with added fishmeal and seaweed.

The results are as follows:

Organic Soybeans	Replications, Yields		
Treatment 1	67.5	67	65.5
Treatment 2	68.75	66	69.5
Control	65.8	69.5	66.8

As you can see from the yield data, no response was measured. This treatment cost over $20.00 per acre plus our work and application costs. We certainly did not get an economic return. This does not mean that the research was a waste of time or that these products do not work. We need to consider the variables. Maybe on other farms with less soil fertility different results would have been obtained. Maybe we applied the treatments at the wrong time. Perhaps wonderful soil biological changes occurred which we did not measure. Maybe the crop quality was greatly improved and again we were not able to measure this change in quality. Maybe nothing happened because these additions were not a limiting factor for this year's crop. The year the study was conducted, we had adequate moisture. In addition, we had a high-yielding crop that was not under a great deal of stress. Another year, things could be different. So we certainly can't say this type of treatment is *useless*. We can only say that under the conditions in which we did the study and during the year in which it was conducted, no yield advantage was measured.

My final conclusion is that we must get the soil healthy and loaded with lots of minerals because this is what will feed the crop. Take all research with a grain of salt. Make sure you see what you are looking at. Maybe the treatment showed positive results, but you weren't able to measure them. Maybe the treatment would have done wonders with just a little more calcium, or less nitrogen, etc. Maybe the treatment needs more time.

Always looking for answers, jumping from one miracle product or practice to another puts you in a rut. Farm and live from the questions, "How can I do this better; how can I make more profit?" If you always question things and live from the questions, many answers will come your way.

INDEX

banded fertilizer, 137
beneficial insects, 278
beneficial organisms, 22, 229, 283-294
Bio-Ag Learning Center, case study, 171
bio-energy, 297-301
Bio-Root, 193, 196
BioCal, 18, 111, 122, 127, 128, 129, 153, 157, 166, 168, 173, 186, 192, 193, 197-198, 199, 200, 201, 216, 220, 221, 226; and pest control, 227
biological agriculture, 9-10, 321-324
biological farmer, 2
biological farming, xii-xiii, 1, 5, 10, 141; consultants, 323; and crops, 12; definition of, 15-16; getting started, 15-23; improvements from, 194-195, 201; and livestock, 12-13; objectives of, 11, 48-50; results of, 13-14; requirements for, 322-323; rules for, 16-22, 324; and soils, 11-12; and weed pressure, 266
biological farms, case studies of, 189-203; negative practices, 162; positive practices, 162-163
biological products, 230-242
biological program, 110-113, 215; creation of, 161-188
biologically grown feed, and consumption levels, 289
biologicals, 229-242
biotechnology, 29, 206
Bird, C., 299
bone meal, 150
borax, 160
boron, 160, 220; beneficial functions of, 109; frits, 160
Bray tests, 101, 108
Brix, 96, 188, 222
broadcast fertilizer, 137
Bromfield, Louis, 169, 189
Bt, 237
bulking, 225
burned lime, 126

calcite, 125, 154
calcitic limestone, 125, 154
calcium, 107, 110, 117-118, 122-124, 125, 129, 152-155, 168, 177; and alfalfa, 218-220; availability, 118; beneficial functions of, 109; correc-

tions for, 182; levels in soil, 153; and potatoes, 224-226; soluble, 218, 220, 224, sources of, 154-155
calcium borate, 160
calcium carbonate, 121, 125, 154
calcium hydroxide, 126
calcium magnesium carbonate, 125, 154, 156
calcium nitrate, 147, 155
calcium oxide, 126
calcium silicate, 126
calcium sulfate, 127, 154, 158
Callahan, Philip, 280, 300
Canadian Journal of Botany, 300
Canadian Journal of Plant Science, 285, 288, 289, 298
Carlson, Dan, 300
catalysts, 232
Cation Exchange Capacity, *see* CEC
CEC, 58-61, 103, 105, 178; and enzymes, 233; increasing levels of, 113-115; and profits, 114-115; and soil type, 115
cells, energy within, 297; health of, 315-316
Chaetomium, 237
chelating agents, and microbes, 68
chemical weed control, 313
chemicals, and earthworm populations, 66-67
chloride, 4
chlorine, 108
clay particles, 56-57
colloidal phosphate, 149
colloids, 42-43
Communications in Soil Science and Plant Analysis, 282
compaction, 247-249; reducing, 249
complete soil test, 98
compost, 147; and humic acid, 240
composting, 257-259; tips for, 258-259
conservation tillage, 246
conventional agriculture, 8-10; compared to biological, 9-14
Cook, R. James, 282
copper, 160
copper chelate, 160
copper sulfate, 160
corn, 205-215; fertilizer practices, 210-215; and nitrogen use, 211-215;

135; and nutrient content of plants, 289; options for, 137-139; organic, 134, 165-167; for potatoes, 223-228; and nutrients, 131-132; for small grains, 217-218; and soil pH, 119-120; quantities, 133; salt, 134; solubility of, 136; soluble, 174; for soybeans, 179-189, 216; starter, 138-139, 179; unacceptable sources of, 19-20; use and placement of, 312; and weeds, 269
fertilizer use, examples of, 177-187
fish, 239-240
fish emulsion, 147
fly ash, 126
foliar application, enzymes, 234
foliar feeding, 229-242
foliar fertilizer, 137
food selection, and insects, 279-281
Forage Fertilization, 292
forages, 218-223; high quality, 313
formazan test, 97
Fraenkel, G., 290
fulvic acid, 240
fungi, 61, 132, 236
gibberellins, 234
glauconite, 152
grains, small, 216-218
granite dust, 152
grass tetany, 151, 175, 221
grazing system, controlled, 313
green manure, 148, 255-257; plants for, 257; ways to use, 256
greensand, 152
growth regulators, 234, 229-242
Gunter, Chris, 226
gypsum, 18, 122, 127, 129, 149, 154, 182

Hancock Agricultural Research Station, 225
hand weeding, 275
hard rock phosphate, 149
harrow, 274
harvest, and soil, 222; and weather, 222
health, 315-318; factors affecting, 318; measurement of, 317-318
herbicides, 20-21, 171, 265; as back-up, 275-276; use of, 313
herbivores, as weed control, 275

high-calcium lime, 18, 125, 129, 154
Hilgardia, 285
hilling, 273
hormones, 234
Horticulture Science, 298
Howard, Sir Albert, 286
humates, 240-242
humic acid, 223, 240-242; benefits of, 240-241; liquid, 241
humus, 21-22, 46-48, 104, 113, 114-115, 251, 257; low level, 248; particles, 56-58; and weed control, 268
humus test, 114
hydrated lime, 126
hydrogen ions, 118-120
hydroxyl ions, 119-120

Idaho phosphate, 149
in-row fertilizer, 137
indicator species, 262
induced resistance in plants, 284
industrial by-products, as fertilizer, 154
infrared light, and plant health, 287
infrared wavelengths, 280
inhibitory sugars, 290
inoculant, 216, 236
insecticides, 313
insects, beneficial, 278; and infrared sensors, 280; and preferred foods, 279-281
ions, 56
iron, 160; beneficial functions of, 109
iron chelate, 160
iron sulfate, 160

Journal of Agricultural Research, 285, 290
Journal of Cellular and Comparative Physiology, 290
Journal of Economic Entomology, 292
Journal of Experimental Biology, 292
Journal of Experimental Zoology, 290
Journal of Insect Physiology, 285

K-mag, 152, 156, 158, 221
Kalium potash, 151-152
Kalmes Farms, 190-195
Kelman, Arthur, 224
kelp, 152, 177, 227, 238-239; benefits

soil health, 316
soil life, 22, 26; and crop nourishment, 132
soil magnesium, 155
soil nutrients, 2; interactions of, 8-82; ratios of, 124-125; reserves of, 77-80
soil particles, 42-43
soil pH, 31-32
soil pores, 43
soil recommendations, 212
soil report, 103-116; reading of, 104-109
soil scorecard, 325-334
soil soup, 173-176
soil structure, 31, 44; and fertilizer, 283; and weed control, 267-268
soil temperature, 33
soil tests, case studies, 163-188; defined, 88-89; differences in, 100-101; and field variability, 99; results of, 103-115; sampling for, 92-93, 99; and seasonal variability, 99-100; and soil chemistry, 100; types of, 88, 91-101; variables in, 98-99
soil water, 44-46, 53, 243
Soil Science, 283, 298
soil-correcting mix, 166-167
soil-plant system, 25-37
Solubor, 160
sound, and nutrient absorption in plants, 300-301; and plant stimulation and growth, 300-301
sound waves, 300
Soviet Plant Physiology, 298
soybeans, 216
Steve Hooley Farm, 198-199
stray voltage, 200
Streptomyces, 237
stress, effect on plants, 285-287
subsoiling, 243-244
sugar content, and plant health, 288-289
sul-po-mag, 152, 156, 158, 221
sulfate of potash-magnesia, 152, 156, 158
sulfur, 108, 127, 129, 156-159, 177; and alfalfa, 219; beneficial functions of, 109; and forages, 218; and ratio to other nutrients, 157; sources of, 158-159
superphosphate, 149, 154, 158

sustainability, 2

Tawfik, Ahmed, 225
temperature, and crop yield, 30
threadworms, 62
tillage, 21-22, 207-208, 243-250, 314; commandments for, 246; improper, 245; types of, 243-245
Tillage in Transition, 246
Tompkins, P., 299
top-dressed fertilizer, 137
trace elements, 19, 108-109, 129, 159-160, 175, 227; and alfalfa, 219; corrections for, 182; oxide form, 20; sources of, 160
trace minerals, 176
transpiration, 52-53
Trichoderma, 237
triple superphosphate, 150, 155
Turner, Newman, 320

ulmic acid, 240
unslaked lime, 126
urea, 20, 134, 146

Vidhyasekaran, P., 293
vitamins, 234

water, uptake by plants, 52-61
Water Deficits and Plant Growth, 286
weather, and crop yield, 29-31
weeders, 274
weeds, 16, 312-313; biological control of, 276; burning of, 274; control principles, 261-276; and crop yield, 33-34; during transition to organic, 308; facts about, 262-263; mechanical control of, 273-275; and mowing, 274; non-toxic control of, 266-276
Willis, Harold, vii, xii
wind, and crop yield, 31
wood ashes, 152

xylem, 53

yield increases, and cultivation, 274

Zimmer, Gary, ix, 191
zinc, 160; beneficial functions of, 109
zinc chelate, 160
zinc sulfate, 160

Acres U.S.A. — books are just the beginning!

Farmers and gardeners around the world are learning to grow bountiful crops profitably — without risking their own health and destroying the fertility of the soil. *Acres U.S.A.* can show you how. If you want to be on the cutting edge of organic and sustainable growing technologies, techniques, markets, news, analysis and trends, look to *Acres*

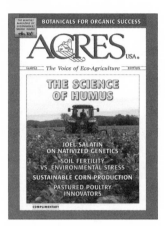

U.S.A. For almost 40 years, we've been the independent voice for eco-agriculture. Each monthly issue is packed with practical, hands-on information you can put to work on your farm, bringing solutions to your most pressing problems. Get the advice consultants charge thousands for . . .

- Fertility management
- Non-chemical weed & insect control
- Specialty crops & marketing
- Grazing, composting, natural veterinary care
- Soil's link to human & animal health

For a free sample copy or to subscribe, visit us online at
www.acresusa.com
or call toll-free in the U.S. and Canada
1-800-355-5313
Outside U.S. & Canada call (512) 892-4400
fax (512) 892-4448 • info@acresusa.com